Samuel Morton PETO
(1809–1889)

Fig.1
Peto in the early 1850s
Illustrated London News 8/2/1851

Samuel Morton PETO

(1809–1889)

The Achievements and Failings of a Great Railway Developer

John G. Cox

Edited and with a Foreword by David Brooke

RAILWAY & CANAL HISTORICAL SOCIETY

*To my parents, Sydney Garth Cox and Elsie May Cox
(née Harbour), neither of whom unfortunately lived long enough
to see the completion of this work.*

First published 2008
by the Railway & Canal Historical Society

www.rchs.org.uk

The Railway & Canal Historical Society was founded in 1954 and incorporated in 1967.
It is a company (No.922300) limited by guarantee and registered in England as a charity (No.256047)
Registered office: 3 West Court, West Street, Oxford OX2 0NP

ISBN 978 0 901461 56 8

Designed and typeset by
Malcolm Preskett
Printed in England by
Biddles Ltd, King's Lynn, Norfolk

Contents

Foreword

by David Brooke, MA, PhD

Formerly Lecturer, Centre for Study of the History of Technology,
University of Bath

Few of those who embarked upon railway contracting in the 1830s, commanded resources in terms of finance and family experience in general building, comparable with those available to Samuel Morton Peto. Fewer still could claim that their first contribution to the nation's track was a feat of construction equal to his Wharncliffe Viaduct. Others struggled through a long apprenticeship in subcontracting or even began as mere navvies. But thirty years later the most spectacular and publicised failure of any British contractor in the nineteenth century, brought his building career virtually to an end and left his reputation irreparably damaged. The intervening period, however, had been one of outstanding achievement and the list of his achievements in Appendix A of this book illustrates that they bear comparison with those of Thomas Brassey, as compiled by Arthur Helps in his life of the great contractor.

John Cox has written more than a very readable survey of Peto's work in railways, docks, etc., at home and abroad. Refusing to accept traditional interpretations, he critically examines this contractor's reputation as a model employer of railway labour and, taking a wider purview, shows that, although the immediate cause of his collapse was a general commercial crisis, all the ingredients for disaster in the form of questionable financial methods existed long before 1866. His business was badly shaken by losses from the Grand Trunk Railway of Canada but the decisive blow came with the hugely expensive Metropolitan Extensions of the London, Chatham & Dover Company; engagement in these proved the final contracting and financial temptation for an increasingly reckless Peto. Brassey, possessing an innate caution and superior judgment, withdrew from involvement with this scheme at an early point in its transition from the East Kent Railway. The lengthy enquiries by the Bankruptcy Court and Peto's own religious denomination, which revealed the extent of his contribution to the collapse of 1866 through incompetence, over-confidence and financial artifice, are given a uniquely comprehensive and judicious evaluation.

He was, says Cox, 'in the vanguard of a new breed of large-scale railway contractors' but the 'ultimate verdict … must be that he was a very successful railway builder and a singularly unsuccessful railway financier.' Samuel Morton Peto has finally found his biographer.

AUTHOR'S ACKNOWLEDGMENTS

THE AUTHOR is indebted to several members of the Peto family, especially the late Sir Francis Peto, for access to family papers, including the notes Peto dictated to his son not long before his death, which formed the basis of the 1893 biography; also for permission to reproduce a number of illustrations in the family collection. Others who assisted included John Bennett, who read the manuscript and made some useful suggestions, Göran Anderberg, who translated some of Danish source material, and of course David Brooke, who besides providing guidance and encouragement during the work's long gestation period, kindly agreed to edit the completed version on behalf of the Railway & Canal Historical Society.

The author is also most grateful to Mrs Kate Petrie for permission to reproduce her late husband's painting of the Wharncliffe Viaduct.

A study of this sort could not have been written without access to a large number of libraries and archival records and the assistance of all their staffs is gratefully acknowledged. Whilst they are too many to be named individually, special thanks are due to the former British Transport Historical Records at Porchester Road, London; National Archives at Kew; and the Institution of Civil Engineers. The author is also grateful to those who have researched relevant topics and published the results in books and journal articles. Most, if not all, are acknowledged in the bibliography.

He is also indebted to Malcolm Preskett who undertook the design and production of the publication on behalf of the R&CHS and to Richard Dean for his considered expertise and advice in producing the maps showing Peto's routes in England.

Preface

THERE is no in-depth modern account of Samuel Morton Peto, which the eminent railway historian, the late Jack Simmons, regretted on at least one occasion.[1] The valedictory account by Peto's eldest son, Henry, came too soon after his father's death and paid too much regard to family susceptibilities, to be truly objective, especially in respect to his failings. He also lacked access to the documentary sources that are now available, especially for the railway companies which employed Peto. This book nevertheless provides an invaluable basis for modern research, as well as some illuminating comments. Without it, any author of a modern study would find it very difficult to make a start.

Until fairly recently the only other writings on Peto have taken the form of journal articles, sections in books on related topics and the briefest of lives. In 1996 however, Edward Brooks, the former Rector of Somerleyton, Peto's estate in Suffolk, published the first full-length biography for more than a century. Whilst this revealed much about Peto's East Anglian links and gave an outline of his railway undertakings, it did not attempt to analyse the relationships between company and contractor, or offer a critique of Peto's business methods.

Despite his final fall from grace, Peto is too important a figure in the Victorian business world and the history of railways in Britain and abroad, to justify such neglect. This monograph is an attempt to redress the situation. It grew out of a detailed study of the Southampton & Dorchester Railway, one of the companies whose lines Peto had built in the hectic Mania period.[2] It draws heavily upon the archival records of all the English, and a few of the foreign companies, which employed Peto over a forty-year period. This was inevitable in the absence of any surviving records of the firm.

A&GWR	Atlantic & Great Western Railroad
A&SLR	Atlantic & St Lawrence Railroad
B&GR	Birmingham & Gloucester Railway
B&OJR	Birmingham & Oxford Junction Railway
C&HR	Chester & Holyhead Railway
CMR	Cornwall Minerals Railway
CP&SLR	Crystal Palace & South London Railway
ECR	Eastern Counties Railway
ESR	East Suffolk Railway
EKR	East Kent Railway
ELR	East Lincolnshire Railway
E&NAR	European & North American Railway
E&PR	Ely & Peterborough Railway
GER	Great Eastern Railway
GNR	Great Northern Railway
GSR	Great Southern Railway of Argentina
GTRC	Grand Trunk Railway of Canada
GWR	Great Western Railway
HR&GR	Hereford, Ross & Gloucester Railway
L&BR	London & Blackwall Railway
LC&DR	London, Chatham & Dover Railway
L&SWR	London & South Western Railway
LT&SR	London, Tilbury & Southend Railway
MDR	Metropolitan District Railway
MKR	Mid Kent Railway
MLR	Mold Railway
MR	Metropolitan Railway
MS&LR	Manchester, Sheffield & Lincolnshire Railway
NR	Norfolk Railway
N&EER	Northern & Eastern Extension Railway
OW&WR	Oxford, Worcester & Wolverhampton Railway
RDR	Royal Danish Railway
RNR	Royal Norwegian Railway
S&DR	Southampton & Dorchester Railway
S&HR	Shrewsbury & Hereford Railway
SER	South Eastern Railway
SL&AR	St Lawrence & Atlantic Railroad
SLR	South London Railway
SVR	Severn Valley Railway
WEL&CPR	West End of London & Crystal Palace Railway
Y&NR	Yarmouth & Norwich Railway

Introduction

ONCE the principle of steam power had been successfully applied to railways, the construction of the system was the joint creation of the engineers, who designed the lines and supervised their construction, the capitalists who provided the considerable sums of money needed and the labourers who carried out the physical work on the ground. Of these three groups, the engineers provided the most glamorous figures, such as the Stephensons and Brunel, as well as other, perhaps equally heroic, if less lauded, practitioners. The capitalists are a more shadowy group, with more than their share of charlatans, to offset the Quakers and other worthies who occupied seats in many railway company board rooms. They were nevertheless essential for the success of any project in the *laissez faire* conditions that prevailed in Britain and most of the rest of the world in the nineteenth century. The labourers were numerically by far the largest group and absolutely vital for the success of any project, although their efforts were largely taken for granted and their achievements passed unnoticed, unless there was a major accident, an outbreak of cholera on site, or perhaps a strike. They were almost totally left to be controlled by their immediate employers, the contractors, who undertook responsibility for the execution of the works under the the the overall supervision of the company's engineer.

The contractors were a diverse group of men coming from a great variety of backgrounds but in the early days of railway building, mostly small-scale operators, often of humble origins, with limited technical knowledge and modest capital resources. Much of the construction work carried out up to and including the 1840s, was in the hands of such men, who were only one step removed from the gangers and navvies they employed. There were exceptions however, most noteworthy of whom was the subject of this study. Unlike some other contractors who grew in stature and resources as they went along, Samuel Morton Peto brought the resources and skills of a large well-established London building firm to the business, when it was still at a comparatively early stage in its development. In this, as in many other aspects of his life, Peto was a man before his time. It was appropriate that he should be in the vanguard of a new breed of large-scale railway contractors, as he was himself a comparatively new man on the building scene, having inherited a large business at the age of twenty-one, and just twenty-five when he obtained his first railway work, for the mighty Hanwell Viaduct.

Peto was in every way well placed to execute a major railway undertaking like Hanwell Viaduct in the mid 1830s. His firm had already carried out not dissimilar building works in London and had access to the necessary plant and materials, as well as labour, both skilled and unskilled. It also had expertise in organising technical operations on a large scale and the capability to supervise a sizeable workforce. The efficient management of labour was to be the keystone of Peto's early success as a railway contractor and as the scale of his operations increased, he perfected a clearly defined hierarchy of command. Although probably not differing greatly from that used by other large-scale contractors, notably Brassey *(plate 15e)*, it must have been fine tuned to have achieved what it did in the Mania years. With himself at the top, and much involved in the projects, at least in the early years, this ran down through agents, who were in overall charge on the ground, subcontractors, timekeepers and gangers, to the skilled and unskilled labourers who carried out the work.

Peto was a superb self publicist and used every opportunity to put a favourable gloss on all his activities, particulary in the way he cared for his workforce. Mainly as a result of the evidence he gave to the Parliamentary Select Committee on Railway Labourers in 1846, he acquired the reputation at the time and ever since, for being a humane and caring employer. The evidence for this is far from overwhelming however, and the subject of Peto, the employer, is urgently in need of re-assessment. This must take account however of the changes in social and economic attitudes, which have occurred since his time. He must indeed be judged as a player on the mid-Victorian scene rather than a man of our time.

Not satisfied with business success, Peto launched himself at quite an early stage on a political career, joining Robert Stephenson, Joseph Locke and George Hudson as part of the railway interest in the House of Commons in 1847. This was a very unlikely achievement for a contractor and demonstrates his enhanced status in the railway hierachy by this time. It also reflected his wealth, which had been sufficient by the Mania years to enable him to purchase a country estate. Membership of the Westminster Club allowed him, in addition to furthering his railway interests, to champion other causes dear to his heart, notably sanitary reform and the right of nonconformists to be buried in Anglican churchyards. This and his Whig-turned-Liberal credentials, have led to a link being made between him and the Radical causes of the period, but examination of the evidence suggests otherwise. Although Peto certainly helped enhance the status of his adopted nonconformity, he is more significant politically as a member of the increasing important mercantile middle classes, and the railway interest in particular.

Another matter requiring scrutiny are Peto's financial activities. He acquired shares at a very early stage in several of the companies for which he worked, sometimes through necessity, as they came to rely upon his credit to complete their projects. He also appears to have developed an early desire to speculate in a range of stocks and to adopt the pose of a wide-ranging entrepreneur. On the railway front, the scale of his financial involvement enabled him before long to obtain company directorships, and also to move into into the world of railway promotion. By the 1850s, he had become as much a railway promoter and financier, as a contractor. Although he always retained his proclaimed primary objective to build railways, he was increasingly leaving its more mundane aspects, such as the supervision of the works, to his perhaps over-indulgent partner, Edward Ladd Betts *(plate15b)*. In the absence of a strong restraining influence, Peto's overambition and reckless speculation not only nearly ruined both of them, but at the same time almost bankrupted a major railway company.

Peto dramatically lost his position in the commercial world during the financial crisis of 1866, leaving it in an atmosphere of financial scandal. Although he managed to extricate himself from the London, Chatham & Dover debâcle and avoided the final disgrace of bankruptcy, sufficient damage had been done to destroy his reputation at the time and seemingly for posterity. Whilst the full intricacies of the affair can never be completely unravelled, it was a sufficiently momentous event in railway history alone, to justify a re-examination of the facts that are known. As for the final verdict on the man himself, the scale of his activities alone justifies the claim that, despite undoubted flaws, he was one of the most important builders of railways in Britain of all time, as well as a major force on the world scene.

ONE

Foundations for Success
1809–1834

Although the future Sir Morton Peto did not inherit any great estate, he could claim to have been born on one – his father, William Peto being at the time a tenant of Sutton Place, the magnificent Tudor domain in western Surrey. Whitmoor House, where he was born on 4 August 1809, although modest by comparison, was still a substantial seventeenth-century farmhouse, its Jacobean architecture, perhaps not entirely coincidentally, being the style chosen by Peto forty years later for his own mansion at Somerleyton. His father was one of the largest farmers in the locality and consequently a person of some standing in the community. Although entirely agricultural, the district was by no means remote by early nineteenth-century standards; the gated road which passed by the farmhouse, joined the main turnpike to London less than a mile away, whilst canal boats plied along the Wey Navigation only a few yards from the farm. Waterways had already greatly improved transport over much of England and were now at the zenith of their prosperity. Their position had yet to be seriously challenged by railways, which were still largely confined to the coal mining areas of the North, although at the other end of the county, the Surrey Iron Railroad, even if only used by horse-drawn traffic, was an indication of things to come. Nearly thirty years would pass, however, between Peto's birth and the building of the first steam railway into the area, the London and Southampton line which reached Woking in 1838.

Samuel Morton was the eldest of the three sons born to William and his wife Sophia, and the only one to have a second Christian name, which he was sufficiently fond of throughout his life to almost always sign himself 'S. Morton Peto'. He received this in honour of his god-father, Thomas Morton, who had been Secretary of the East India Company in the late eighteenth century. It is tempting to see this significant figure in the commercial world of his day as the role model adopted by Peto during his own lifetime; which is quite likely, as he endeavoured to perpetuate the link by bestowing the same second name upon all of his own sons. There was certainly a strong family link with the activities of the East India Company, Peto's father's sister, Ann, having married Thomas de la Garde Grissell, another prominent figure in the Company. He would in due course be the father of Peto's future business partner, Thomas Grissell. Another member of the Peto's family who had moved to London and done well for himself, was his father's brother, Henry. Already the owner of substantial building business, he was destined to play a very significant part in our subject's early life.

Before Peto was seven the family had moved to Buckinghamshire and his early education was at a school in Marlow run by a Mr Field. When he was twelve, however, almost certainly at his uncle Henry's instigation, he was sent as a boarder to the Brixton Hill Academy in London, which was run by Mr Jardine, who was a Congregational Minister. Henry Peto, although an Anglican like the rest of the family at the time, was one of the school's principal benefactors. Most of the pupils were either the sons of tradesmen or destined to follow such a calling. Even at this tender age, Henry may have intended that his nephew should join his firm. Henry also strongly approved of the school's evangelical religious stance, which meant the boys received copious instruction on the scriptures, although there was also a sound grounding in mathematics and English and some attention paid to the classics but little to modern languages, something Peto would come to regret when obliged to travel abroad on business in later years. His capacity for speedy calculation, mastery of spoken and written English, as well as copper-plate handwriting all clearly testify to the quality of the education he received at the school, which was in the tradition of the dissenting academies of the previously century. Besides displaying all-round ability as a pupil, Peto also excelled at drawing, the lessons in this subject at school preparing him for the training he was subsequently to receive as a draughtsman.

There never appears to have been any intention on the part of his parents for Peto to return to the farm. With no land for him to inherit and the depressed state of agriculture in the period following the end of the Napoleonic wars, a trade in town appeared to have better prospects, especially when Uncle Henry offered the lad an apprenticeship when he left school. Even at this stage there was the prospect of a significant place in the firm eventually, as Henry had no children and may have already been looking to the next generation of the family for successors to carry on the business after his death. So it was that in 1823, at the age of fourteen, Peto embarked upon a seven-year apprenticeship in the building trade.

Whatever intentions Henry Peto may have had ultimately for his nephew, he bestowed few favours upon him in the early years of his apprenticeship, requiring him to work long hours and perform all the routine labours demanded of the other apprentices. In addition he was expected to study architecture and draughtsmanship at evening classes; which he enjoyed, especially when there were opportunities to visit historic buildings and indulge in his hobby of sketching. The work most of the time was severely practical, however, the aim being a thorough mastery of all the skills of the building trade, which presented few problems for this intelligent and hard-working young

man. In later years Peto was to look back with some pride on the three years he had spent at the carpenter's bench and how he had proved himself a fully-fledged bricklayer by laying 800 bricks in a day. His efforts and proven competency were rewarded by being given charge of work on a number of sites, including the construction of a house the firm was building in Carlton House Terrace for the politician and wit Horace Twiss.

In 1830, when Peto was twenty-one and only just out of his articles, an event occurred which transformed his prospects overnight. His uncle died after a very short illness, bequeathing his entire business equally between Peto and another nephew, Thomas Grissell, who was eight years older and already a partner in the firm. The other members of the family, including several as closely related to Henry as the favoured nephews, were only left comparatively small sums, which caused considerable resentment and a threat on their part to contest the will in the courts. They certainly had prima-facie grounds to object to the settlement, as the will had been drawn up only days before Henry's death, with just the two nephews, a witness and the lawyer present. There are grounds to suspect that the two main beneficiaries put pressure upon their dying uncle for their own ends. Peto subsequently claimed that his uncle's sole desire had been to ensure that the firm remained intact and in capable hands; his enemies in later years might have drawn a parallel with another railway entrepreneur, the notorious George Hudson, whose start in life also owed much to a legacy, which some alleged, had been obtained in questionable circumstances.

The inheritance was not without its problems, however, as the estate was heavily mortgaged and a substantial claim for compensation still remained outstanding in connection with the contract for the New Custom House in Lower Thames Street which had partially collapsed during its construction due to faulty foundations. The ensuing litigation between builder and architect had severely damaged Henry's reputation. Although his nephews eventually settled the matter, the memory of the case, which for Peto went back to his early years with the firm, could well account for his lifelong reluctance to resort to the courts to settle any dispute.[3]

At first the new Grissell & Peto partnership continued with the variety of regular work, such as house building and the repair of fire damage, upon which Henry had built its prosperity; although, as has been seen, he had not been averse to undertaking the occasional large-scale work and assuming the then fairly novel role of 'building contractor.' This meant agreeing an overall price for a building and then engaging sub-contractors for the work he did not carry out himself. This system had been pioneered early in the century by Thomas Cubitt, who had proved it to be more economical than the time-honoured practice of an architect engaging different master craftsmen for the various types of work involved. It was slow to be adopted in a tradition-bound industry, however, and although offering a potentially high return to the contractor, brought additional responsibilities and consider-ably more risk. Despite their uncle's experience with the Custom House, within two years of taking over the firm, Peto and his partner under-took a number of substantial building contracts.

The first came as early as 1832, when the contract was obtained to rebuild the old Hunger-ford produce market in the Strand (plate 2b), to the designs of the architect Charles Fowler, for a company headed by the Earl of Devon. Although at £42,000 their tender was the lowest, the directors were at first inclined to reject it, mainly because Peto when he presented it, cut such a youthful figure, being only twenty-three at the time. He was subsequently summoned to attend a meeting of the full Board of the company, which gave him his first opportunity to display his talents as an interviewee, having no difficulty in assuaging the doubts of even the most hardheaded of his interrogators. He even enlivened the proceedings with a touch of humour, saying he would willingly take to wearing spectacles to look older, and if that was not good enough, could call in his partner 'who looked old enough for anything'. The work was satisfactorily completed and a handsome profit made. In the next three years further contracts were obtained for two West London churches, St John's, Paddington and Holy Trinity, Sloane

Street and two theatres, the Lyceum (plate 2a) and the St James's, the latter being completed in just four months, despite serious difficulties over access to the site. The firm also did some of the masonry for the Thames road tunnel at Rotherhithe, which was being constructed by Marc Brunel and his son Isambard.

Nor were the partners' efforts limited to the metropolis. Besides some housing in Brighton, which went back to their uncle's time, in 1833 they had obtained a major contract in Birming-ham, to rebuild the King Edward VI Grammar School in New Street, which was worth over £30,000. In successfully executing this building to the designs of the up-and-coming architect, Charles Barry, they demonstrated their ability to execute work in the Gothic as well as the Classical style, which, together with the connec-tion forged with Barry, paved the way for the subsequent Houses of Parliament contract. It was in Birmingham also, that Peto met John Thomas, who was following a much humbler calling at the time as a stone mason, but would, with the assistance of Barry and Peto become a leading Victorian sculptor, as well as the architect of Somerleyton Hall.

In 1831 Peto had consolidated his partnership with Grissell by marrying his eldest sister, Mary. Such kinship ties were a feature of many business arrangements of this sort, partnerships being the most common form of enterprise in the era before the advent of the modern joint stock company. The couple spent their honey-moon in the Cotswolds and it was then that Peto probably had his first experience of steam-powered travel. This was on the Cheltenham & Gloucester Rail Road, which normally used horses but was experimenting in 1831 with a steam locomotive, *The Royal William*. It was not a good introduction to the new mode of travel, however, as the engine broke down before they reached their destination and the journey had to be completed by coach. Peto's desire to make the journey in the first place, and on his honeymoon as well, suggests he may already have been seriously considering becoming involved in railway building, only a year after George Stephenson had completed the Liverpool and Manchester line.

TWO

Early Railway Contracts 1835–1843

THE GREAT WESTERN RAILWAY

BY THE early 1830s several projects were under way to link London by rail with the main provincial centres, one of the most important being the line to Bristol promoted by the Great Western Railway (GWR). This scheme finally obtained the approval of Parliament in 1834 and the company's engineer, Isambard Brunel, promptly drew up the specifications for the various contracts to build the railway. These were for sections of track on an average less than ten miles in length and major works, such as viaducts and tunnels. Grissell & Peto were amongst the first firms tendering for the contracts between London and Reading, attracted by the potentially lucrative profits and the affinities of some aspects of the new works with the firm's building operations. It was not their first attempt to obtain a railway contract, having tendered slightly earlier for work with the London & Birmingham Railway, including the line from Euston to Camden.

Peto tendered initially for two major works on the GWR, the viaduct over the Brent valley at Hanwell and the bridge across the Thames at Maidenhead. Both were formidable undertakings, estimated by Brunel to cost £36,000 and £32,000 respectively. Peto was awarded the first with a tender of £39,487 but his price for the second was nearly £2,000 more than the successful bid. He also failed to obtain the contract for the line between Iver and Maidenhead, the only section of track he tendered for at this stage, his price being nearly

double that of the successful contractor, James Bedborough, and also considerably above Brunel's estimate. Assuming the engineer was not wildly out in his calculations, Peto had shown his capacity to estimate accurately the cost of bridges and viaducts but displayed an inability to match this when it came to earth-moving and the laying of permanent way. This is not altogether surprising, as he had had no experience up to that time of the latter, whilst the former bore some relation to his previous building work. When on unfamiliar ground, unlike some of his competitors, he had wisely erred on the side of caution. Bedborough obtained the Iver contract at well below Brunel's estimate but curiously enough this was accepted, whilst a bid from another contractor considerably lower than Peto's for the Hanwell Viaduct, was rejected.

Hanwell Viaduct (plates 3, 4, 5) was the first work of any significance on the line west of the capital and therefore particularly attractive to a London-based firm such as Grissell & Peto. It was also one of the most prestigious on the whole GWR, and recognised as such by the directors, when they decided to name it after Lord Wharncliffe, the peer who had played an important part in guiding their Bill through the House of Lords. In due course his coat of arms would be displayed on the southern face of the viaduct, where, despite the subsequent widening, it remains today. Designed by Brunel the elegant structure was 300 yards long, its eight arches each having a span of 70 feet and a maximum height of 65 feet above the level of the river.

Although work had begun promptly on the viaduct, it was not long before Peto was at loggerheads with the company's Resident Engineer, J. W. Hammond, who adopted a high-handed attitude towards contractors, demanding amongst other things that Peto use red facing bricks instead of the yellow variety that were more readily available locally. He complied initially, putting himself to the additional expense of transporting the required bricks from Clapton on the other side of the city. But when Hammond further demanded that the mortar used should match the colour of the bricks, he devised a subtle scheme to get the better of him. Not averse to playing practical jokes, he ordered his foreman to have the workmen be seen mixing cochineal into the mortar when Brunel was next on site and to say that this was being done on Hammond's instructions. Brunel duly fell for the ploy, ordering 'the foolery to be stopped'. Although this no doubt gave Peto considerable pleasure at the time, Hammond may well have had his revenge, as Brunel, in a letter which survives, tells Peto that his foreman had 'in one instance privately, or by trick' evaded his orders and demands that he be dismissed. If the cochineal incident was the reason, Peto appears, not for the last time, to have been prepared to pass blame onto a subordinate. He also appears in part at least to have managed to elude the red bricks requirement, as looked at today most of the viaduct appears to have been executed either in yellow brick or a mixture of red and yellow.

Otherwise work progressed well, despite wet weather in the spring of 1836 and a blockage on the Grand Junction Canal, which caused a shortage of sand. Brunel was able to inform the GWR Board in May that the viaduct was in a forward state, adding that there had been 'strict and constant supervision' of the work. It was finished by the summer of 1837, by which time Peto had been given an additional contract for the nearby skew bridge over the Uxbridge Road (*plate 6a*), a fairly small but demanding undertaking, costing £6,570. A more substantial contract followed later the same year for the five miles of track between Iver and Hayes. This was worth £35,967 and Peto's first taste of this type of work. It was also the first railway contract he obtained without competition, perhaps because by this time Brunel was under pressure from the company to open the line between London and Maidenhead as soon as possible, and wanted to avoid any unnecessary preliminary delays. Peto was clearly a contractor in whom he had confidence to execute the work as rapidly as possible. Nevertheless there were some more acrimonious exchanges between the two before the task was completed, Brunel insisting that the foundations of the bridge over the Grand Union Canal were replaced and accusing Peto on another occasion of lack of effort, which was probably unfair, because the company minutes record that work was being carried on here day and night.

The GWR proceeded to engage Grissell & Peto to erect the station buildings at the Paddington terminus. This contract was not executed for a fixed sum like the other works he had undertaken for the company, but rather on a 'schedule of prices'. This arrangement had much in common with the traditional method of charging for buildings prior to Cubitt's innovations, the materials used being specified and charged for separately, as were the wages of the different types of operatives employed. It was a system Peto would have liked, as it relieved him of the commitment to an overall price and made the company responsible for seeing that both materials and labour were used economically. The original station at Paddington was only a temporary wooden structure, and consequently easy to erect and not expensive. Peto did have some problems however, with the road bridge outside the terminus, which was included in the contract, Brunel being dissatisfied with the materials used and some of the workmanship, and much of it had to be rebuilt.

Although Peto had not originally tendered for any contracts west of Maidenhead, the opening of the railway there in 1838 and the imminent completion of portions beyond, encouraged him to undertake contracts further from his London base. The same year the GWR engaged him to build the section of line from Tilehurst, just west of Reading, to Goring, which included Basildon Bridge over the Thames. This was a much more demanding task than the Hayes line, which only crossed the flat Middlesex countryside; as it required deep cuttings through the steep-sided

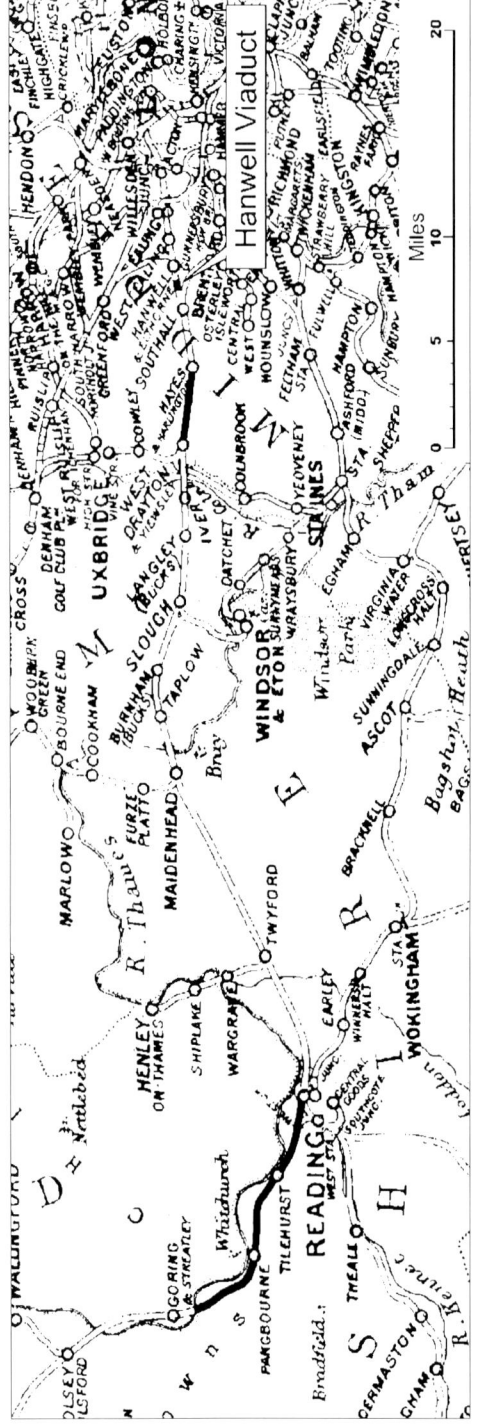

bluffs of chalk that flanked the southern bank of the Thames, as well as including a substantial brick viaduct across the Thames just east of Goring. Peto also obtained this contract without competition, although he was obliged to agree to a fixed price of £68,795 for the six miles of track and the bridge. Despite inclement weather and some obstruction from the owner of Basildon House, who tried to get the work stopped whilst he was in residence, he completed the contract in less than eighteen months.

The firm also erected the station buildings at Reading, which Brunel dismissed as 'a paltry design.' It was whilst he was supervising the final stages of work here on 24 March 1840 that Grissell had a very lucky escape, when the nearly-completed structure felt the full force of an exceptional violent gust of wind. A large section of the roof was torn down and several of the men working there were injured, one fatally, whilst bricks and debris were scattered over the forecourt. Grissell received a severe wound on the head from one of the bricks dislodged from the chimney and was carried unconscious to the George Inn, where he received medical attention.[4] Although he made a complete recovery, this traumatic experience could have made him disinclined to engage further in the railway side of the business, the financial risks of which were already causing him many a sleepless night. He soon handed over entire responsibility for this to Peto, concentrating his own efforts after 1840 on the execution of the most important building contract the firm had yet obtained, for the new Houses of Parliament.

There were some grounds for Grissell's fears. The early GWR works had certainly not been particularly profitable, due mainly to the intense

Fig.2: **The eastern end of the Great Western Railway, sections of which – highlighted by the solid line – were built by Peto: namely the Hanwell Viaduct and the bridge over the Uxbridge Road about a ¼ mile to the west, the section of track from Hayes to West Drayton and that from Reading to Goring which included Basildon Bridge over the Thames west of Pangbourne.**

competition to obtain them and the low profit margin before any allowance was made for unforeseeable factors, such as an increase in the cost of materials or labour. One of the other firms employed on the London to Reading section certainly reneged on his contract and another, David McIntosh, ended up in a long-runnning dispute with the company over the payments due to him. Peto was involved in a similar dispute, the sum involved amounting in the end to £160,000, which made it necessary for him to borrow £25,000 to keep going. He was left with the option of resorting to the courts or submitting his claim to Brunel, a clause in the contract not unusual at the time, specifying that the Chief Engineer would be sole arbitrator in such cases. It was a difficult decision, especially as his relations with Brunel had been strained and Grissell refused to advise him on the matter. Persuaded perhaps by his uncle's experience of a lengthy and expensive lawsuit, he chose the latter course. It proved the right decision, as Brunel speedily settled most of the outstanding bills, Peto receiving nearly all he had claimed. McIntosh probably wished he had followed the same course, as his court case dragged on for many years, his grandson eventually receiving the payment, something *The Times* would remind its readers of during Peto's dispute with the London, Chatham & Dover Railway a quarter of a century later.[5]

Despite the favourable arbitration verdict, Peto did not get on well with Brunel describing him as 'very overbearing and inconsiderate'. Years later he recalled one instance of the great engineer's behaviour, to his son, Henry:

> Most of these meetings took place at night after dinner – Brunel smoking all the while. Once an appointment was made for 7 o'clock; towards midnight the servants came in evidently anxious to shut up the house but Father, whose patience was exhausted, had them tell Brunel he was still waiting. Brunel came in with many apologies and declared he had quite forgotten the engagement but arranged another interview.[6]

Although Peto tried to obtain the much more substantial contract for the Temple Meads terminus station at the other end of the line at Bristol, no further contracts were forthcomiing from the GWR after he had completed Reading station. Whilst working in the town, however, he seized the opportunity to indulge in a bold private speculation. Amongst the properties his uncle owned had been Furnival's Inn in Holborn and this may have given him the idea of building an hotel opposite the station in Reading to accommodate the new clientele the railway would bring. Named appropriately *The Great Western (plates 7a & b)*, it was completed within three years of the railway reaching the town and was initially let to the lessee of the refreshment rooms on the station. The speculation proved a great success and the hotel was to cater for travellers for more than a century.[7]

BIRMINGHAM STATION

BEFORE completing the last of his undertakings for the GWR, Peto had started work on the erection of Curzon Street station in Birmingham *(plate 8b)*, at the end of the railway from London, which had been opened in 1838. Unlike the makeshift wooden buildings at Paddington, this was a very significant work, designed, like the magnificent Doric Arch, which fronted the company's London terminus at Euston, by the distinguished neoclassical architect Philip Hardwick. The directors of the London & Birmingham Railway had engaged Thomas Cubitt, the greatest building contractor of the day to erect the Euston Arch; so to be entrusted with the terminus at the other end of the line was a significant achievement for Peto and more than adequate consolation for his failure at Bristol. Hardwick was responsible for the main station building, which cost £26,000, the most notable feature of which was the Greek Revival façade with its four Ionic columns. This had some affinities with the Lyceum Theatre in London and consequently an architectural form with which Peto was familiar. The 270ft-long train shed behind, although costing less, probably represented a greater technical challenge, as its roof was an advanced design, making extensive use of wrought iron and glass.[8]

The only serious problem Peto encountered at Curzon Street was connected with the façade of the building rather than the train shed; although the difficulties did not arise on the site but many miles away on the Cromford Canal, which was being used to transport the massive stone slabs required for the columns, from a quarry in Derbyshire. Whilst in transit one of these became stuck in a tunnel, bringing navigation to a halt and a consequent threat of litigation from the canal company. Peto received news of this at his London home late in the evening and without delay took the night train north to assume charge of the operation to clear the tunnel. His success in doing this was a personal triumph and an illustration of the challenges faced by the early railway contractors; the speed with which he reacted was significantly assisted in this instance by the facilities offered by the recently completed railway. The station was duly completed in 1839.

THE LONDON & BLACKWALL RAILWAY

H AVING extended his operations well beyond his London base, Peto tendered in 1838 for two sections of track on the Manchester & Birmingham Railway, as well as the viaduct at Congleton on the same line, but in both cases was underpriced by other firms. He was also unsuccessful for the same reason in his bid for two sections of the North Midland Railway and the stations at Leicester and Derby on the Midland Counties line. His efforts to obtain railway works nearer home with the Greenwich, Eastern Counties and Brighton companies, also failed, and the same fate befell his tender for the Birmingham company's Dunstable branch. These setbacks were probably as much a result of his cautious approach to railway contracting as the intensely competitive conditions prevailing at the time. There was some consolation for this frustration when the London & Blackwall Railway (L&BR) engaged him in 1839 to complete part of its line. He had applied for the contract earlier but had been undercut by Robert & George Webb, whose

price had been nearly £10,000 less than his, and also below the estimated costing that had been prepared for the company by its Assistant Engineer, George Parker Bidder. Peto's price had also been below this, although only by £300, which would probably have made the work unprofitable, as Bidder had already acquired a reputation as a calculating genius (plate 15c).

The L&BR regretted its decision to engage the Webbs before work had proceeded very far. Realising they were losing money they asked to be released from their contract, leaving the company and its Chief Engineer, Robert Stephenson, in the embarrassing situation of having to call in other contractors to complete the railway. In the end the remaining work was divided between Grissell & Peto and Thomas Jackson. Employing two firms appears to have been deliberate policy on the company's part, as Peto did not compete for the work that was placed with Jackson. In February 1839 the L&BR Chairman confidently predicted:

> When the energies of two contractors with ample resources shall come into operation, the works will be so acclerated as to leave no doubt of their being successfully terminated by the time specified.[9]

Peto had been requested by the company to apply for the new contract, so had no inhibitions in seeking preferential terms. The main concession he obtained was to charge for work on a schedule of prices. This offered the prospect of a higher profit, besides being better suited to a partially completed undertaking, than the fixed-price alternative. An exception however was the building that housed one of the fixed steam engines which powered the system of ropes used to propel the trains, for which a price of £10,482 was agreed at the start.

Stephenson had opted for fixed engines in preference to locomotives, because they were at the time better suited to the steep gradients and attaching ropes to the trains was practical on a line less than three miles in length. It was nevertheless an expensive undertaking, as, in addition to the cost of the fixed engines, the track had to be carried nearly all the way on brick viaducts. The company ran into financial difficulties before the line was finished and although Peto received nearly £55,000 for the work he carried out, he was obliged on several occasions to accept payment in L&BR shares. This was the first occasion he was required to assist a company in this way and a foretaste of much that was to occur elsewhere later. The railway was completed in July 1840.

THE SOUTH EASTERN RAILWAY

THE L&BR was not alone in having difficulties raising funds – the depressed state of the investment market in the late 'thirties making it difficult for other companies to complete their ongoing projects. One such was the South Eastern Railway (SER), which was engaged at the time on the construction of its main line from Redhill to Dover. Although the contracts on the difficult eastern section from Ashford to Dover had been let in 1839 and some work had been done between Folkestone and Dover, two years would pass before Peto could start in earnest on his major undertaking for the company.

This was the viaduct over the Foord Valley just east of Folkestone, which, although 50 yards shorter than that at Hanwell, was an equally demanding undertaking, being twenty feet higher and requiring nineteen brick arches (plate 9). Helped no doubt by his previous experience, Peto managed to complete this major work in less than two years without any significant difficulties – in fact, there was not a single fatality reported during its entire construction. At the same time he had also managed to complete the nearby Martello Tunnel and a section of permanent way near Hythe, as well as the stations at Folkestone and Westenhanger. The SER appears to have paid for all his work on a schedule of prices, his total payments exceeding £200,000. He could hardly therefore object to having to accept £20,000 of this in company bonds. The railway was finally completed through to Dover in February 1844.

Although the SER invited Peto to tender for proposed branches to Canterbury, Ramsgate and Margate, in the end these contracts were placed with other firms. His only other works for the company were the short extension to a new London terminus at Bricklayers Arms, relaying the track on the Canterbury–Whitstable line and widening the Maidstone branch. He also erected the Royal Pavilion Hotel at Folkestone for the company. The Bricklayers Arms line was constructed on a timber viaduct and completed in a few weeks in 1844. The hotel, despite its grand name, was a converted boat builder's barn and architecturally undistinguished.

During the course of working for the SER, Peto again made a number of significant contacts. These included two members of the Cubitt family, William, who was the company's Chief Engineer and his son Joseph, the Assistant Engineer; and his future partner, Edward Ladd Betts. The SER employed Betts to ballast the whole of the Redhill–Ashford line and to construct a length of track near Hythe. This adjoined Peto's works, which were on the other side of Saltwood tunnel, and when the contractor responsible for the tunnel proved unsatisfactory, the company asked Peto and Betts to complete it together. In the course of this operation the two became friends, Betts becoming a frequent visitor to the house Peto was renting at Hythe in order to be close to the works. There he made the acquaintance of Peto's younger sister, Ann, who had come down from London to keep house for him. Betts married Ann soon shortly after the SER works were completed.

The SER also had romantic associations for Peto. His first wife, Mary, had died in May 1842 and it was during the celebrations marking the opening of the second section of the line, from Tonbridge to Ashford the following November, that he met his second wife, Sarah, who attended the event with her father, Henry Kelsall. Besides being the owner of a large textile mill in Rochdale, he was a major railway investor, part of the Lancashire interest which played such an important part in the financing of the early schemes. So for one such as Peto, with aspirations to rise within the railway hierarchy, he could be said to have chosen well.

THE YARMOUTH &
NORWICH RAILWAY

PETO had hardly got back to London before his involvement in the Yarmouth & Norwich Railway (Y&NR) made it necessary for him to move up to East Anglia. Promoted by Robert Stephenson and his father George, this scheme was an attempt to partially revive an earlier project for a trunk line from Norfolk to London. Given the relatively depressed state of the investment market in the early 1840s, however, even a comparatively small undertaking such as this, could not easily raise all the capital it needed. So it seems likely that the Stephensons, no doubt aware of the assistance which Peto had recently provided the Blackwall company, approached him as a possible backer. He may have also been personally attracted to the venture on account of his firm's business connections with Yarmouth which went back to Henry Peto's time.

In due course Peto became the Y&NR's Resident Engineer, leasing Bracondale, a substantial property on the Carrow side of Norwich while he worked from an office at St Michael-at-Plea in the city. Most of the actual construction was probably carried out by George Merrett, an experienced railway surveyor, and a friend of George Parker Bidder. Bidder was Stephenson's assistant again on the Y&NR, making it in terms of personalities in some respects a re-run of the L&BR. Peto was certainly happy to continue his collaboration with Robert Stephenson, whom he found much easier to get on with than Brunel, describing him as 'a true gentleman, who always considers the needs of others'. Peto had also established a close working relationship by now with Bidder, so was at no disadvantage when he took over the post of Y&NR engineer from Stephenson during the course of the construction of the line.

Although the Y&NR crossed flat country most of the way, at the Yarmouth end it was necessary to traverse part of the Norfolk Broads, where the track had to be laid on marshy estuarine silt and peat. The Stephensons had previous experience of such work on the Liverpool & Manchester line, whilst for Peto crossing such terrain would be valuable training for the operations he would soon carry out in the Fens. Just outside Norwich, there was also the not inconsiderable task of diverting the course of the river Yare for about half a mile.

Although it is not possible to know how much assistance Peto provided, or what form this took, the directors held him in sufficient esteem to ask him to preside at the dinner held at the Assembly Rooms in Norwich in April 1844 to celebrate the opening of the line. This event occurred against a background of considerable criticism of the way in which the directors had handled the company's affairs during the construction of the line, some sections of the railway press having attacked them for refusing to divulge their deliberations to the outside world, in effect acting as some sort of cabal. Enemies of Peto in later years would have seen this as the first instance of his sinister hand at work but it is not clear how much influence he really had in the Y&NR Boardroom. If he had been largely responsible for the success of the venture, Peto should have redeemed his initial outlay with a handsome profit on the works, as in the end the railway cost nearly £10,000 per mile, considerably in excess of Robert Stephenson's original estimate, which was only £7,000.

On the day of the opening, a special train conveying invited guests made a successful round trip from Norwich to Yarmouth. Another excursion laid on just prior to the main event for the benefit of the Bishop of Norwich, who was apparently an early railway enthusiast, was somewhat more eventful, as Peto recalled many years later in a letter to Bidder's son.

As we were travelling with the Bishop & your father's staff in 3 ballast trucks with temporary seats we saw a cow on the line. I was driving the engine, your father, as he did sometimes … was standing with me on the foot board smoking. I said to him: 'What shall I do?' To pull up without contact was impossible. He said: 'Go at her!' I put on the steam and she disappeared over the side ditch … At lunch the Bishop delighted with his ride … said to me: '

What was the white thing I saw fly over the ditch?' I said: ' We shall see, My Lord, on our return journey.' Going back we found the dead cow twenty feet from the outer rail in the marsh … The Bishop then realized how much he owed under Divine Providence to your father's advice, for, had the contact been after slowing down the engine, we should most likely have gone into the side cutting filled with water. [10]

In the ten years between obtaining the first GWR contract and completing the Y&NR, Peto had built nearly fifty miles of railway. Although a fairly modest mileage compared with what he was to construct later, this had included two very large viaducts and a major station building, whilst the total value of the works was about half a million pounds. Railway contracts during this period were still mostly small, so Grissell & Peto could claim not only to be in the top league of contractors; but by extending operations from its original London base into the Home Counties, the Midlands and finally East Anglia, had demonstrated their ability to operate nationwide. The firm had also obtained, and successfully completed, some of the largest works on offer. With this reputation and the resources to assist in the funding of new projects, its services would almost certainly be sought by other companies, when the next wave of expansion in the rail system took place.

Meanwhile there had been no diminution in the firm's traditional building work. This reached its high point of activity in the early 'forties, when, besides the first stage of the new Houses of Parliament, there were major contracts for other public works in London, including Nelson's Column (plate 8a), an extension to the Royal Dockyard at Woolwich, the Model Prison at Clerkenwell and a number of club houses, including the Reform in Pall Mall, of which Peto himself would soon become a member. Outside the metropolis important improvements were carried out to the navigation of the river Severn, as well as some additional locks on the Grand Union Canal. But the great days of canal building in England were over and, whatever Grissell may have thought, Peto now had no doubt that the future for him lay with railways.

THREE

Mania Bonanza
1844–1847

THE EASTERN COUNTIES AND NORFOLK RAILWAYS

THE Yarmouth & Norwich project had come at an opportune moment for Peto. By the time he had completed the line the investment market was beginning to revive, setting in train the process which was to culminate in the 'Railway Mania' of 1844–7. The changed economic climate not only made it easier than it might have been, for him to dispose of any shares he still held in the company, but also encouraged the promotion of two new schemes, which taken together resurrected the earlier plan for a trunk line from Norfolk to London. One was put forward by the Y&NR itself, soon to be re-named the Norfolk Railway (NR), for an extension from Norwich to Brandon; the other, promoted initially by the Northern & Eastern Extension Railway (N&EER), would provide a link southwards from Brandon, through Ely and Cambridge, to connect with its existing line from Newport (Essex) to London *(plate 10)*. Before the scheme was finalised the N&EER had been absorbed by the Eastern Counties Railway (ECR), which, although it had just completed a line from London to Colchester, saw the Brandon route as the best means of tapping the potentially lucrative Norwich traffic.

The interests of both the ECR and the NR appeared to be best served by employing the same contractor to build both the new lines and Peto, with his Y&NR connection, was the obvious person to carry out the task. It may also have helped that the engineers for both schemes were his friends and recent collaborators, Robert Stephenson being in charge of the ECR part of the project and Bidder the NR portion. So, apparently without having to compete with any other firm for either contract, Peto embarked early in 1844 upon the construction of eighty miles of new track, the longest he had so far undertaken.

Peto had already carried out some work for the ECR, having been engaged the previous year to erect a number of the stations at the eastern end of the Colchester line and to convert the railway from the unusual 5ft gauge upon which it had been built, to the soon to be pronounced standard 4ft 8½in. Colchester remained a terminus for some time and this must have been one reason why Peto decided to build a hotel adjacent to the station. Known as the *Royal Victoria (plate 11b)*, like its predecessor at Reading, this was a private speculation on his part, although significantly larger. Lewis Cubitt, the future architect of King's Cross station in London, designed this grandiose Italianate building at a cost of £15,000. Unlike Reading, it was a commercial failure from the start, being too far from the town and not receiving sufficient patronage from railway passengers . It had ceased to function as a hotel by 1850 and ended its days as the county lunatic asylum.

Before work commenced on the Newport–Brandon line, Peto undertook a five-mile-long branch from Broxbourne to Ware and Hertford for the N&EER. Although this presented no significant physical difficulties, it did involve him in an acrimonious dispute between the company and the Cheshunt Turnpike Trust, over the road crossing to be provided at Ware. Under the terms of its Act, the company had been required to construct a viaduct here, but this had not been completed by the time the railway was ready for use. After the Trust refused to come to a short-term arrangement, Peto took matters into his own hands, proceeding to lay the rails across the road by stealth over one weekend. When the Trust threatened litigation, he told them that he would remove the rails and provide a viaduct within a fortnight, provided they gave an undertaking not to object to the completed work, saying that if they did, he would relay the rails at once to allow the trains to cross. His ultimatum was accepted, and the Trust had to content itself for the time being, with a basic wooden structure. Peto celebrated the success of this ploy when the railway opened, by providing a sumptuous dinner at Broxbourne for the directors and officials of the company, who would soon be in no doubt that their contractor, besides being a man of action, also possessed more than his fair share of guile. When Stephenson subsequently required Peto to replace the wooden structure with a proper bridge at his own expense, he did not object, declaring it was his desire to comply with the Chief Engineer's wishes at all times. When the bridge was subsequently completed, however, it was found to be narrower than originally specified, allowing Peto to recoup some of his extra expenditure. The incidents at Ware tell us a lot about Peto's *modus operandi* at an early stage in his railway contracting career.

The following year Peto began work on the Newport–Brandon line, which by this time had been taken over by the ECR. He had agreed to build this for a fixed sum of £10,000 per mile exclusive of stations and other embellishments. But there appears to have been a lack of clarity in some of the specifications however as, early on, he was in dispute with the company as to whether he was required to lay single or double track between Cambridge and Ely. Who was responsible for this astonishing situation is not clear, but in the end he was obliged to provide the additional set of rails without any extra payment.

Peto completed the Newport–Brandon line in July 1845, within eighteen months of starting work, earning himself a bonus of £25,000, which was some compensation for the additional expenditure on the bridge and the double track. The construction had been assisted by the comparatively favourable terrain, the only

significant work being a short tunnel at Audley End, which had only been needed to placate the owner of the nearby stately home. Peto was also engaged to carry out a number of additional works, the most important being the stations at Cambridge, Ely and Peterborough. Unlike the line, he charged for the stations on a schedule of prices and their cost reflected this. Stephenson had originally estimated that the station at Cambridge (*plate 11a*) could be erected for £10,000 but the ECR eventually paid more than five times that sum, whilst the buildings at Ely and Peterborough cost £80,000 and £90,000 respectively. Although some other work was included, such as sidings and warehouses, the final price in both cases appears excessive, especially as, with the exception of Cambridge, the buildings were modest structures, if designed to a high standard by the company's architects, Sancton Wood and Francis Thompson.

Before the main line was completed, Peto had been engaged to build an important extension from Ely to Peterborough, to link up with the London & Birmingham Railway's line from Blisworth. Originally an independent concern, the Ely & Peterborough Railway [E&PR] had been absorbed by the ECR by the time it was finished in 1847. Peto also built branches from Cambridge to St Ives, and from March, on the E&PR, to Wisbech and to St Ives. With the exception of the thirteen miles from Cambridge to St Ives, which were built for a fixed sum of £160,000, the new lines appear to have been charged for, either on a schedule of prices, a per mile basis, or a combination of the two.

The E&PR was the most important addition to the ECR system, as it provided a link with lines leading to the Midlands and North. Its construction, however, necessitated traversing part of the Fens, which, although largely drained by this time, was still not the easiest of terrain. The line also had to cross the New Bedford River, the artificial drainage channel which carried water into the Wash. The company entrusted Peto with the task of negotiating with the Bedford River Commissioners over the form this bridge was to take and he succeeded in persuading them to agree to a structure £6,000 cheaper than that originally

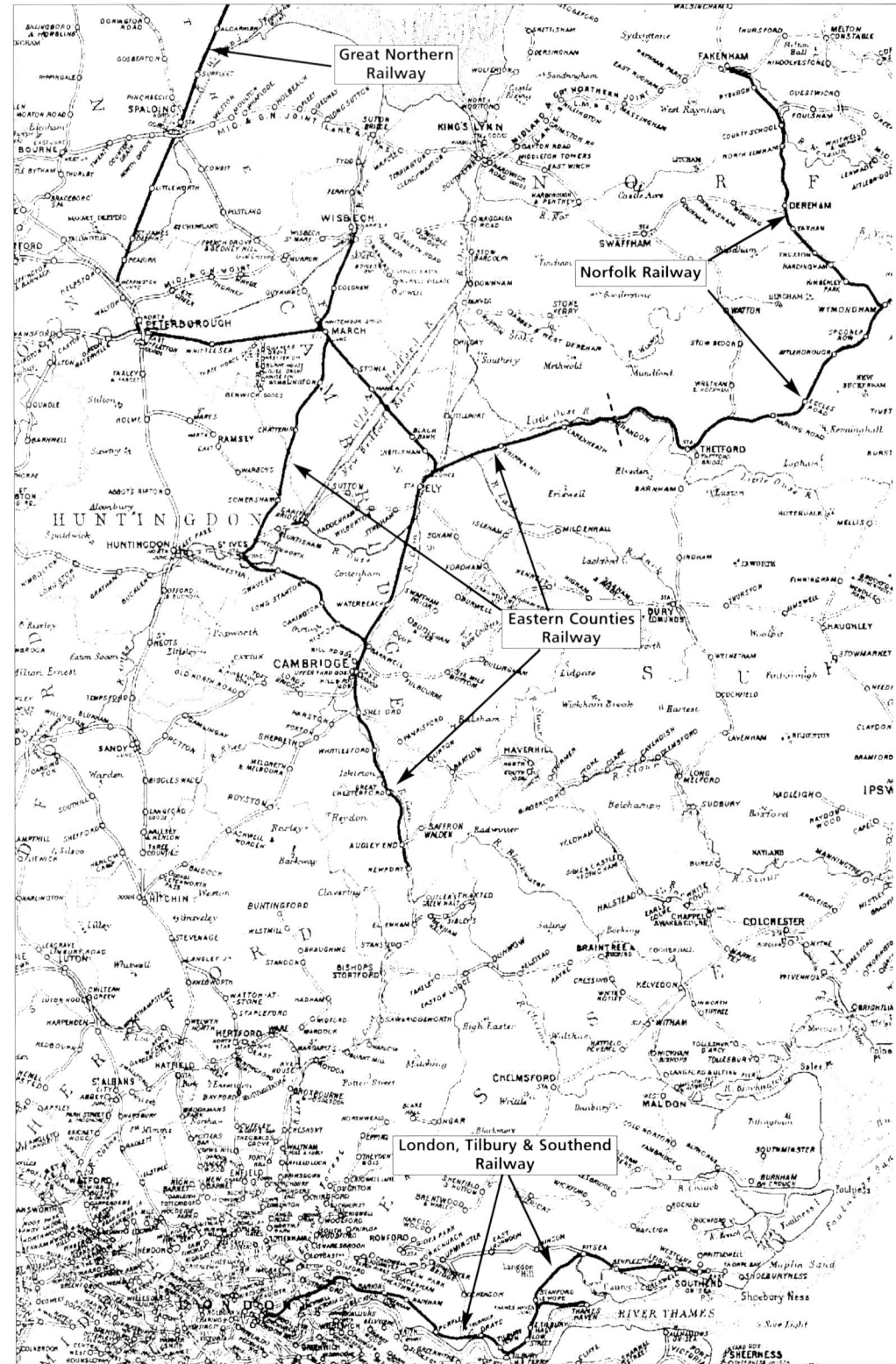

Great Northern
Railway

Norfolk Railway

Eastern Counties
Railway

London, Tilbury & Southend
Railway

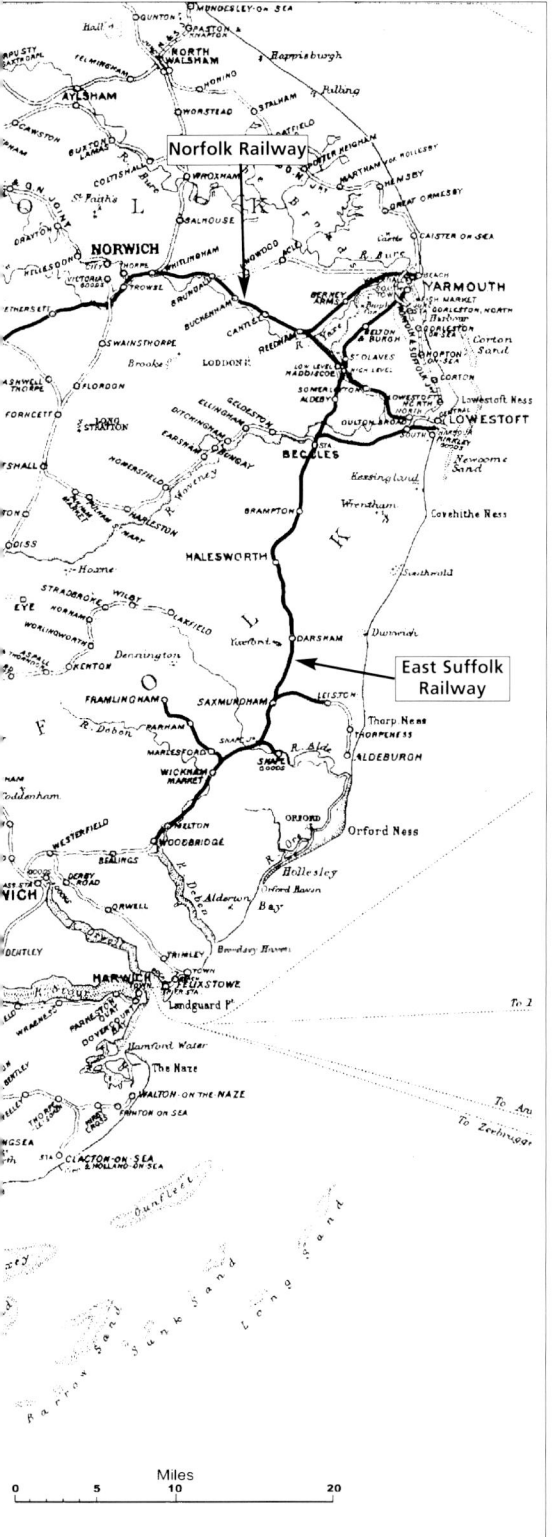

proposed. The resulting wooden trestle structure was 120 feet in length, the largest work of this kind built in the country up to that time. Its construction was sufficient of an achievement for Captain Coddington, the Board of Trade Inspector, to include these words of praise for the contractor in his report sanctioning the opening of the railway:

> The energy and skill shown by Mr Peto, the contractor, who has executed the work, reflect high credit upon his talent and engineering skill.[11]

Peto claimed he was paid in all £1,123,000 by the ECR. As his firm's account books have not survived, it is not possible to say how much profit he made on what was then his most expensive railway undertaking. Nearly all of the work was carried out before 1849, when the ECR had to admit it was virtually bankrupt. The company had been in serious financial straits at least eighteen months earlier when Peto still had work in hand. In May 1846 he had attended a Board meeting to press his claim for £140,000 that was due to due to him, but was only paid £20,000 of this in cash. The company's only way out was to meet most of what was due to him through bills and bonds, £80,000 being paid in this way in 1846, £415,000 in 1847 and £192,000 in 1848. Even this was only possible due to the timely intervention of George Hudson, who not only underwrote the company's losses but proceeded, by highly dubious means, to raise the additional capital needed for the ongoing works. Although in no way involved personally in Hudson's malpractices, which included the falsification of the company's books to allow capital to appear as revenue, Peto undoubtedly benefited indirectly from the activities of the 'Railway King' whilst he

Fig.3: **The rail network in the Eastern Counties with routes built by Peto highlighted with solid lines.**
His routes spread north and east from Newport in Essex but he also built the line from London to Southend-on-Sea.

was Chairman of the ECR. Had he not been at the helm Peto might well have found himself in considerable difficulty; as it turned out the ECR only owed him about £7,000, when Hudson was forced to resign in 1849.

Peto was obliged, however, to appear before a Committee of Investigation set up by the ECR shareholders to look into the company's finances. When he was questioned about the large sums expended on the stations, where he would have been open to criticism, he sought to place blame elsewhere, saying:

The works for the stations were not done by contract but by measurement and valuation … from the plans of Mr Thompson, and afterwards Mr Hunt was called in …
The prices fixed were ten per cent under the Board of Trade prices. It was arranged before the works were commenced that they should be so paid for. Not only was I so paid, but all the other contractors.[12]

The Board of Trade rates referred to, were to be used again by Peto in less auspicious circumstances, to justify his own charges for work on the London, Chatham & Dover line.[13] The Committee nevertheless accepted his evidence on its face value and chose not to pursue the matter. They had bigger more vulnerable witnesses to examine, in particular the fallen 'Railway King'. Such were Peto's enforced investments in the company by this time, that he was able to see his brother-in-law and new partner, Edward Ladd Betts, installed as Chairman.

Peto had also undertaken to build the Norwich–Brandon line for the NR for £10,000 per mile, a comparable rate to that charged its predecessor for the Yarmouth line. Although the extension crossed somewhat more hilly country than the Y&NR, the only work of any significance was the two-span swing bridge 100 feet in length over the river Wensum at Trowse, just outside Norwich *(plates 16a)* This was the first of a number of such structures he would build, to cross navigable waterways in the eastern counties. Designed by Bidder, it was not finished when the rest of the line was opened in July 1845, passengers having to be conveyed to and from Norwich by coach for several weeks.

Although he had been able to complete the rest of the line from Brandon *(plate 12c)* at the same time as the ECR, there is no record of the NR paying Peto any bonus for his success in so effectively co-ordinating work on the two projects.

The only additional works on the line were some new locomotive and carriage sheds at Norwich, needed to handle the additional traffic, and a modest expansion of the station, which included a clock tower that enabled travellers and the locals to set their watches to the new 'railway time' *(plates 12a & b)*. Although this was the same as standard Greenwich time, it was dubbed 'Peto time' during the 1847 election, when he was the Liberal candidate for the city.

Before the Brandon line was completed, the NR, encouraged by the heady atmosphere of the all prevailing Mania, had, like the ECR, embarked upon a number of branches. The first was a line from Wymondham to East Dereham, which Peto was engaged to construct in 1845. The cost of this is not known but when the branch was subsequently extended to Fakenham, the contract price for this was £62,000, which worked out at only just over £5,000 per mile.

In 1845 Peto also obtained the contract for a branch from Reedham, on the former Y&NR, to Lowestoft. This had originally been promoted by an independent company but was soon absorbed by the NR. The sum agreed for its initial construction was £80,000, which worked out at just over £7,000 per mile, a modest cost considering it included two swing bridges, one at Reedham and the other at Somerleyton. Peto had a special interest in the project, having taken the first steps before it was completed, towards transforming Lowestoft from a decayed harbour into a modern port and fashionable watering place. In 1844 he had also purchased the house and estate at Somerleyton, which lay along the route of the railway. His protégé, the former stonemason, John Thomas, was engaged to transform what had been a modest Queen Anne house, into an imposing mansion. Thomas probably also designed the nearby station, which, with its square tower, parapet and latticed windows was much grander than any of the other country stations on the branch. In deference to the new owner of the estate, it displayed Peto's newly acquired coat of arms over the entrance porch (plates 13a & b). Whether he or the company paid the bill for these embellishments is not known.

Both the NR and the ECR were persuaded to invest heavily in the harbour improvements Peto had instigated at Lowestoft, which were designed in part, if not entirely, by Bidder. Most of the work involved was carried out by Lucas Brothers, originally a Norwich firm, with which Peto would maintain a longstanding connection. When he decided shortly after acquiring Somerleyton, to expand brickmaking on the estate, Lucas Brothers took the lease of the works and before long were sending bricks as far afield as London. They also took over his Belvedere Road works in Lambeth and proceeded to establish a large building business in the metropolis.

The Lucases also built most of Lowestoft South Town, the new resort created by Peto on open land facing the sea, beyond what was at the time the silted-up channel between the harbour and Lake Lothing. According to tradition one of his first acts on coming to Somerleyton had been to invite the local Improvement Commissioners to lunch and get them to agree to sell this tract of land to him for £200. When they had agreed, he took them into the drawing room where to their surprise the plans for the proposed new town were laid out. By the early 'fifties the nucleus of the modern holiday resort had been created, with two long terraces of apartment houses, a few superior residences and a hotel. The development was designed to attract a mainly middle-class clientele, one of the entrances of the harbour being converted into a pier with a subscription library. A promenade was also laid out along the seafront. Besides the rewards from the property speculation, Peto no doubt hoped that the scheme would result in more passengers using the railway; an important consideration, as by this time he had acquired a major personal stake in the NR.

Peto's other private ventures at Lowestoft were all closely linked with the railway he had built to the town. Before it was completed, he is said to have boasted that it would be possible to have fish landed there on sale in Manchester the next day, and to this end not only erected a new fish market but acquired a portion of the local fishing fleet. An even more daring undertaking, was the creation of the North of Europe Steam Navigation Company, with a fleet of paddle steamers linking Lowestoft with the Danish port of Tonning. Unfortunately, like the hotel at Colchester, this speculation was misguided and involved him in a loss of about £60,000.

In all Peto constructed seventy-two miles of track for the NR between 1844 and 1849, but the company was in a perilous financial position before all the work had been completed, and badly in debt to him. The collapse of the Mania had depressed the value of the company's shares, whilst disappointing traffic receipts and excessive support for Peto's Lowestoft schemes added to its troubles. The directors were consequently in no position to resist Peto's demand for a place on the Board, although some of their number did hold out against this for a while. He was made a director in 1848, his first elevation to such an office and a portent of the path he was destined to tread. At the same time he had been endeavouring to arrange for a lease by the ECR, which was finally obtained in 1849. He was elected Chairman in 1851, a post he held for four years, during which time he tried unsuccessfully to negotiate an amalgamation of the company with the ECR.

THE SOUTHAMPTON & DORCHESTER RAILWAY

HIS substantial commitments in East Anglia had not prevented Peto from undertaking another contract, far removed geographically from his main sphere of activity at the time. In 1844 he had agreed to construct the Southampton & Dorchester Railway (S&DR), a sixty-mile-long line promoted by Charles Castleman, a Wimborne solicitor, who, after unsuccessfully courting the GWR, had eventually obtained the backing of the London & South Western Railway (L&SWR). The engineer, William Scarth Moorsom, who had been briefed by the S&DR directors to negotiate a fixed price contract, duly told them that he had agreed a price of £480,000, with 'one of the leading firms in the country'. The negotiations

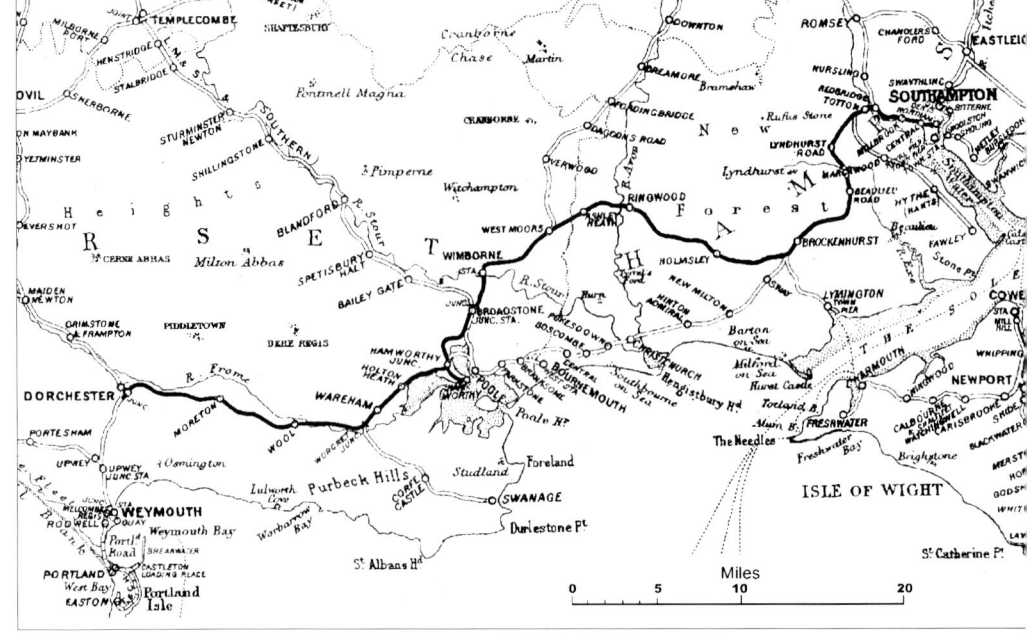

Fig.4: **Peto built the complete route from Southampton to Dorchester highlighted with the solid line**

had been carried out privately with Peto, no other parties apparently been involved. With the services of contractors very much at a premium in this hectic period of railway building, the directors were probably relieved at the outcome.

Peto decided from the start to sub-contract all the works on the line, the first and only time he would do this in Britain. The line was duly split up into a number of sections and each placed with a separate subcontractor, Peto of course deciding who was to carry out the work in each case. No records survive of the prices and terms he negotiated with his sub-contractors but having agreed a fixed overall price with the company, it seems almost certain that he made similar arrangements with the subcontractors. He gave the job of supervising their work and overseeing the whole project to James Beatty, who had worked for him on the Y&NR and was one of his most trusted agents *(plate 15d)*. Peto also agreed to invest £25,000 in the company and when the subscription contract was drawn up prior to the submission of the scheme to Parliament, he

was the second largest investor after William Chaplin, the Chairman of the L&SWR.

Like his East Anglian lines, the S&DR crossed comparatively easy terrain, although the route chosen by Moorsom was so circuitous that it quickly acquired the nickname 'Castleman's Corkscrew'. This did not bother Peto, who was quite satisfied to have access to the harbour facilities at Southampton and Poole, for shipping in most of the materials needed for the construction of the railway. Even before the S&DR had obtained its Act, he had assembled several ship-loads of iron rails at Poole, making the company an advance of £800 to cover the cost.

Although good progress was made on the Dorset portion of the line, the company became involved in arguments with two public bodies, which threatened seriously to delay work at the eastern end. The Town Council at Southampton insisted on the lengthening of the tunnel needed to protect a park being laid out at the northern side of the town; whilst the Commissioners for the Royal Woods and Forests refused to accept

Moorsom's proposed route for the railway through the New Forest. Both disputes must have annoyed Peto at the time, but their resolution in the end enhanced the value of his contract. At the same time as agreeing to lengthen the tunnel, the s&dr directors decided to lay a second set of rails on the first four miles of the line from its junction with the l&swr at Southampton; whilst the route through the New Forest eventually agreed, was two miles longer than had been originally intended. Peto's contract was appropriately amended and he was also promised a bonus of £5,000 if he completed the railway by 1 June 1847.

In addition, there were some engineering difficulties connected with the tunnel at Southampton. This was constructed by the 'cut and cover' method and the unstable nature of the ground made for frequent slips along the sides of the cutting, whilst heavy rainfall flooded the workings several times. The most serious incident occurred however, after work had finished and only a few hours before the first train was due to pass through the tunnel, when a large section of the brick arch suddenly collapsed. Although the damage was soon repaired, this mishap was a considerable embarrassment to all concerned, including the Government Inspector, Captain Coddington, who had officially sanctioned the opening of the railway only the previous week. Peto's response was characteristic. He hurried down to Southampton the next day to inspect the damage, along with Coddington, Moorsom and Beatty. But when the trouble was found to have been caused by Beatty's decision to block up a disused canal tunnel that ran under the new railway tunnel, he had no alternative but to admit liability and foot the bill for the repairs. Beatty was soon forgiven for his error and Peto would subsequently give him sole responsibility of the execution of another major undertaking.[14] Although the blockage in the tunnel disrupted services for some weeks, the company still paid Peto his bonus.

The collapse of the Railway Mania late in 1847 put an end to Peto's hope of further contracts from the s&dr, which had plans for branches to Lymington, Blandford and Weymouth, as well as an ambitious extension westwards to Exeter.

He was fortunate in receiving payment for nearly all his work in cash, although after the l&swr absorbed the company in 1848, he did experience difficulty in obtaining payment of some fairly small outstanding sums that were due to him. The new owners also refused to renew the contract he had been given to maintain the track, when this came up for renewal after three years. Peto was also held responsible for the heavy wear suffered by the track, but this was probably as much a consequence of the many curves on Moorsom's chosen route, as any failings on the contractor's part. In the end the railway had been built for just over £10,000 per mile, comparable with Peto's East Anglian lines of the same date, but it cannot be ascertained whether he increased or reduced his own profit by subcontracting the work. Nor is it known how the subcontractors fared, although there was a report, later denied, that the one responsible for the tunnel at Southampton was in financial difficulties. Peto probably did quite well from his investment in the company, as the terms under which it was acquired by the l&swr, were quite generous.

THE GREAT NORTHERN RAILWAY

MEANWHILE Peto had been able to maintain the momentum of his operations in eastern England by taking an important new contract with the Great Northern Railway (GNR). After a long and expensive battle with George Hudson, this company had obtained an Act in 1846, authorising it to build a trunk line from London to Yorkshire, considerably shorter than the existing Midland route controlled by the 'Railway King'. The GNR also provided a more direct route from London to Peterborough than the line through Ely that Peto had just completed. From Peterborough northwards the GNR proceeded by way of Grantham, Newark and Retford to Doncaster, the so called 'towns line', which it was hoped would in due course be extended to York. In addition it was proposed to construct a 'loop line' from Peterborough, through Boston and Lincoln, to rejoin the main line at Doncaster.

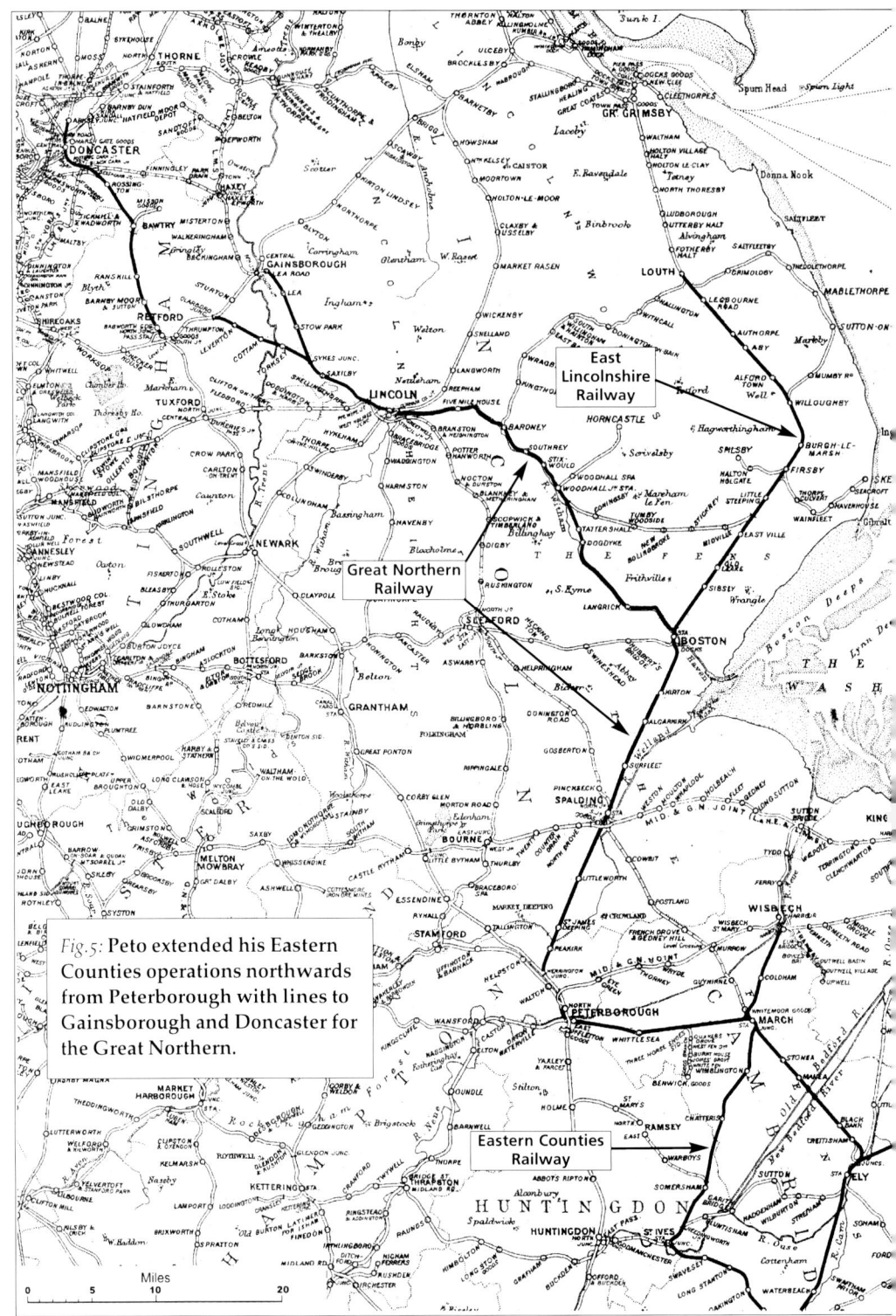

Fig.5: Peto extended his Eastern
Counties operations northwards
from Peterborough with lines to
Gainsborough and Doncaster for
the Great Northern.

East
Lincolnshire
Railway

Great Northern
Railway

Eastern Counties
Railway

Plate 1:
Peto in the early 1840s.
Portrait by John Lucas in the possession of the Peto family.

Peto formed a partnership with his cousin Thomas Grissell when both were in their 20s and soon secured important contracts in London.

Plate 2a:
The rebuilt Lyceum Theatre, London – shown here in 2008.
Author

Plate 2b:
The remodelled Hungerford produce market in the Strand.
(E. Walford, 'Old and New London', popular edition Vol.3, 1897)

Peto's first railway contract was for work on the Great Western Railway to Brunel's designs.

Plate 3a:
Wharncliffe Viaduct at Hanwell.
(J. Bourne: *Great Western Railway*)

Plate 3b:
Close-up of the Wharncliffe Viaduct after widening, with the original pillars erected by Peto on the south (left-hand) side. *Author*

Plate 4: The Wharncliffe Viaduct, named after Lord Wharncliffe who supported the GWR Bill in the House of Lords, captured in a watercolour by J.W. Petrie showing a 'King' class locomotive with Centenary Stock on a pre-war express heading west

Plate 5a: **The arms of Lord Wharncliffe on the southern side of the viaduct.** *Author*

Plate 5b: **View of original (i.e. south facing) side of the Wharncliffe Viaduct in 2008 with a train crossing.** *Author*

Plate 5c: **Close up of some of brickwork of the viaduct suggesting that a mixture of colours may have been used.** *Author*

Other major works on the GWR built by Peto:

Plate 6a: **The bridge over the Uxbridge Road which was demolished in the 1840s after being damaged by fire.**

Plate 6b: **Basildon Bridge over the Thames, west of Pangbourne.**

(both J. Bourne: *Great Western Railway*)

Plate 7a:
Great Western Hotel at Reading in 1908.
By kind permission of Reading Public Library

Plate 7b:
The former Great Western Hotel at Reading, renamed the Malmaison, in 2007.
Author

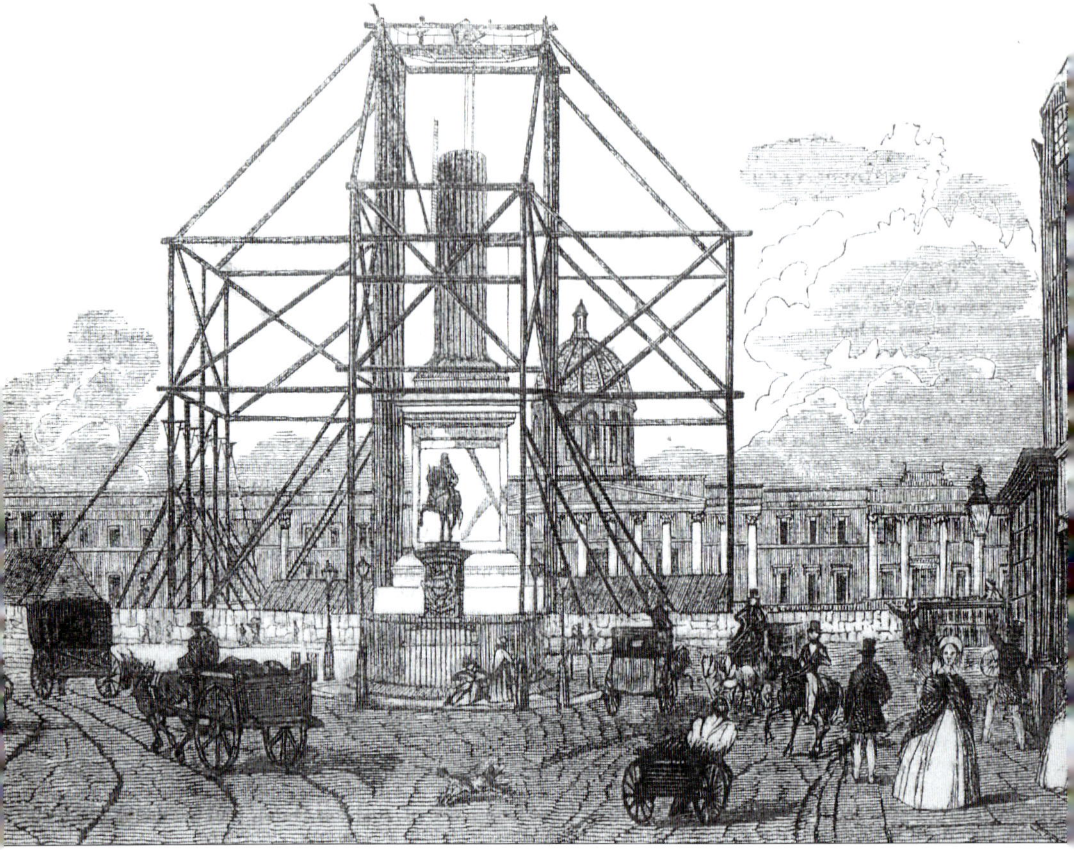

Plate 8a:
The base of Nelson's Column during its construction by Peto showing the scaffolding employed. (*Illustrated London News* 10/9/1842)

Plate 8b:
The front of Curzon Street Station, Birmingham built in 1838 – the trainshed behind was demolished in the 1960s. *Author*

In December 1846 the GNR gave Peto the contract for the entire eighty-six mile long 'loop line', which was subsequently modified to share the Manchester, Sheffield & Lincolnshire Railway (MS&LR) line north of Lincoln and to join the 'towns line' at Retford, rather than Doncaster. At the same time Thomas Brassey was given the London–Peterborough works, the GNR thereby ensuring that the larger part of its works was in the hands of the two greatest railway contractors of the day. Neither appears to have been obliged to compete in the open market for the contract. The remaining works were not let at this stage, although Peto was apparently led to believe he would in due course be given the contract for the extension to York.

Joseph Cubitt, known to Peto from their South Eastern days, was the GNR Chief Engineer, and almost certainly had some influence in the selection of the contractors, as well as the arrangements under which they carried out the work. With the Mania at its height, even a company as large as the GNR, with its substantial financial backing in the City of London, was not in a position to dictate terms to Peto or Brassey, and both were probably satisfied with what they obtained. Each was to be paid on a pre-arranged schedule of prices, which included materials and labour. It was the first time that Peto undertook a whole railway on such terms. The company did however insert a clause into his contract, which obliged him to complete 'the loop' within two years of the start of work, on penalty of a fine of £1,000 for each week he exceeded the agreed deadline. Although the reverse of the bonuses he had received elsewhere, the terms overall were no doubt sufficiently favourable to make the risk worth taking. The only extra remuneration offered to him was a premium of 5% over the basic rates for the erection of some of the more remote stations.

The GNR was feeling the effects of the collapse in the share market by the spring of 1848 and requested both Peto and Brassey to accept debentures in lieu of cash for all monthly payments over £50,000. Both initially refused to comply but Peto relented when it became clear that 'the loop' works would be given priority. Brassey's payments were subsequently limited to £10,000 per month and work consequently slowed down considerably on the London–Peterborough line. Peto on the other hand pressed ahead with vigour on 'the loop', Cubitt reporting to the GNR directors that there were 3,000 labourers, 250 horses and eight locomotives at work there in November 1848. The line had been completed to Retford by late the following year, including a short cut-off line from Sykes Junction to Leverton, with a bridge over the Trent for the MS&LR. In the meantime Peto had taken and completed the contract for the portion of the 'towns line' between Rossington, just south of Doncaster and Retford, although at rates that were 20% below those for 'the loop'. The GNR had by this time however, abandoned the proposed extension to York in favour of arrangements to run its trains along other companies' lines, and the only work he was given north of Doncaster was the two miles section to Askern, where an end-on junction with the Lancashire & Yorkshire Railway gave the GNR access to Leeds.

The rapid construction of 'the loop', as with Peto's other works to date, had been assisted by the easy terrain, the only significant engineering works being the bridges over the Witham at Boston and Bardney, the latter, a wooden structure 729 yards long, taking several months to complete (plates 16b & c). The GNR directors had nevertheless found the construction of the line a drain on the company's resources and were under increasing pressure from the shareholders to avoid unnecessary expenditure. They consequently set up a Lincolnshire Subcommittee to examine the outlay on 'the loop' and when its report appeared, there was much that was highly critical of Peto. The Board recorded in the minutes of its 21 December 1848 meeting that:

> On comparing the schedule of prices for building and excavating with those of the most competent and substantial builders and excavators in the county, they [i.e. the Lincolnshire Committee] find them from 25 to 33 per cent above what are considered profitable rates. But unreasonable as these

rates appear, they by no means describe the whole evil of the principle of letting works on a schedule of prices, which is a direct bounty on extravagant quantities, and the more exorbitant the scale, the more active the inducement which is palpable at every station and work they have been able to examine. The small gate stations are all fitted with platforms and sidings sufficient for the largest stations on the line … cast iron fixed at £2-10s per ton … stuck to in every possible place … treble chairs for the main line of rail … at £12-10s per ton whilst the company's contract for chairs is £5-15s.[15]

Peto appears to have repeated the practices that had earlier made the stations on the ECR so expensive. In his defence it must be said, however, that the differences between the contract rates and those obtainable at the time of the investigation, were between those apertaining at the height of the Mania and what were available in the subsequent depression; also, having once agreed the rates Peto had been committed to them, whatever happened, and the cost of labour and materials could have increased after he accepted the contract. If this had happened the GNR would certainly not have compensated him. It is also unlikely that the local firms whose rates were quoted, could have executed as large a work as 'the loop', built it as quickly as he did, or taken shares rather than cash for any significant portions of their work.

The Committee recommended that the company should not engage Peto to carry out some additional works needed at a number of the stations on 'the loop', because its enquiries had revealed that he had been making a profit of a shilling a day for each workman employed on the buildings at Boston and Lincoln. Its advice to the Board was:

Although the accommodation and alterations … are urgently needed for the business of the company, your Committee cannot feel themselves justified in recommending this being executed under the contract of Messrs. Peto & Betts if it is possible to put them … under a different scale and principle of payment, for, so far

from deriving any advantages … from their experience, architectural and otherwise, in the construction of the stations, … no similar instance of extravagant blunders could be produced in the Kingdom.[16]

Although this was almost certainly an exaggeration, the report probably sealed Peto's fate as far as the outstanding contracts on the 'towns line' south of Retford were concerned, which were placed with other firms. He was, however, allowed to complete the short but important link between 'the loop' and the London line at Peterborough, which included the bridge over the Nene. This proved a difficult undertaking, due to the extensive flooding which occurred in the area during the winter of 1849–50.

As in the case of the ECR, there is no criticism on record of Peto's work from the company's engineer, which is not surprising as his staff had been responsible for supervising the execution of the works. There was more trouble for Peto when the GNR refused to reimburse the dues he had paid for the transport of ballast on the Foss & Witham Navigation, which he was entitled to claim under the terms of his contract. It is some indication of the company's feelings that although the sum involved was less than £2,000, the matter had in the end to go to arbitration.

Although Peto's connection with the GNR ended on a sour note, the company had reason to be grateful for him for building 114 miles of line in less than three years, even if 'the loop' was soon relegated to the status of a branch line. How much he received in payment is not known, although the engineer's original estimate for 'the loop' alone, was well in excess of £1 million. His profits are even more conjectural but would appear to have been considerable for a work on this scale. There was also the significant advantage compared with the ECR and NR contracts, of not being obliged to accept an excessive quantity of shares in payment for the works.

THE EAST LINCOLNSHIRE RAILWAY

THE East Lincolnshire Railway (ELR) had been set up to construct a branch, running from the GNR 'loop' at Boston, to Louth and Grimsby. The company's engineer, John Fowler, had let the original contract for the whole of the line to Waring Brothers. Although this firm made some progress at the northern end, by early 1848 the whole project had proved too demanding for them and Fowler was obliged to find another firm to complete the outstanding work between Louth and Boston. Peto was the obvious choice, as the ELR wanted to complete the line at the same time as 'the loop' and the GNR had not yet called into question his charges for its line. Having received his assurance that the outstanding work would be completed by the following September, Fowler proceeded to engage Peto without bothering to seek any other tenders. He told the ELR directors:

> I entertain little doubt from the well known character of the contractors for punctuality and from the ample and business like preparation they have made for the vigorous prosecution of the works, that the whole [line] will be completed within the time named in the contract.[17]

Warings had contracted to build the section of the line from Grimsby to Louth for £123,000 but it seems likely that Peto charged for his work on a schedule of prices, seeing that he was building 'the loop' on those terms and had insisted on this system when he took over the uncompleted Blackwall line. The ELR agreed to pay him up to £10,000 per month, three-quarters of this in cash, which was comparable with the arrangement he had been obliged to make for his later GNR works.

Peto was set quite a demanding task to complete thirty-three miles of track in nine months, especially as it included nine bridges, the largest being a single arch with a span of twenty feet over the river Ludd. Not only was the task completed on time, but for the second occasion in a year, Peto's works were singled out for praise

by a Board of Trade Inspector, in this case Captain Laffon, who was particularly impressed with the Ludd Viaduct. How much profit Peto made from the ELR contract is not known, but there was certainly no opportunity for expensive additional works; the only significant station along the route, at Louth, perhaps on the advice of the GNR, being entrusted to another firm.

THE REWARDS OF BUSINESS SUCCESS

PETO had reaped very considerable rewards from the Mania. Between 1844 and 1847 he had built over 400 miles of track and embarked upon contracts, worth in all well over £3million. With the exception of the GNR, which was probably beyond the capacity of any one firm at the time, he had been the sole contractor. All the contracts had been placed with him directly by the companies; none, as far as can be known, being opened to competitive tender, a reflection in part of the times, but also an indication of the prestige he had acquired as a contractor. It had been necessary, however, for him to substantially assist two of the companies, the ECR and the NR, the latter becoming a long-term liability.

It might appear surprising, when things were going so well with the railway side of the firm's business, that Peto and Grissell should decide to part company at the height of the Mania in 1846. This had been on the cards for some while, however, given Grissell's dislike of railway contracting and the loosening of their bond of kinship with Mary Peto's death. The partnership was dissolved amicably early in 1846, with Peto taking over nearly all the railway works and Grissell taking control of all the building operations, the most important of which was the Houses of Parliament. After Grissell had finished this he retired to Norbury Park, his country house near Leatherhead, where his anti-railway sentiments would surface again, when his view of the North Downs was threatened by the proposed railway to Dorking.

Peto was not tempted to retire to his country seat, although he almost certainly could have done so on the profits he had earned from the Mania contracts. But he was still only in his thirties and blessed with boundless energy and a strong constitution, whilst his success had given him more self-confidence than ever. But the parting from Grissell brought about a change of outlook and purpose, as is evident in a letter he wrote at the time:

> All my stock in trade will be £25,000 of plant, and all the rest clear capital … being all in money but the £25,000; I shall have it clear and ready for using, or taking on some one or two large railway works, and nothing else, which will only half occupy my time, and then the power of capital will always, with my previous experience, give me a preference without my being known as a competitor to anyone.[18]

He was clearly intending, whilst still remaining a railway contractor, in the future to concentrate his energies on financing the projects with which he was involved, assuming by this means that he would be able to exclude competitors and largely dictate his own terms to the companies involved.

Severing his involvement with the Houses of Parliament contract also enabled Peto to embark upon a political career, which he could not have done as a public works contractor. The 1847 general election took place towards the end of the Railway Mania and Peto's success in being elected as Liberal MP for Norwich, was, along with his recently-acquired country seat, the outward manifestation of the successful entrepreneur. Although money bought influence, he was nevertheless a controversial candidate in a keenly contested election, partly because, since his second marriage he had become a Baptist, and Nonconformists were looked upon with disfavour by many of the Anglican Whigs, who constituted a significant portion of his potential support in the constituency. Although many in the business community admired Peto's success, his opponents did their best to exploit his involvement with the still

far from universally popular railway and attack his strong advocacy of free trade, which was seen by some to threaten the future of agriculture. A hostile poster in the form of a mock address by Peto to the electors of the city endeavoured to exploit these phobias, as well as the religious issue:

> I am for free trade in everything but particularly in railroad shares and I anticipate the greatest advantages will ultimately accrue to all classes of the community in general, and to the contractors in particular, from the railroad extension measures, to which consent has been given in the late sessions of Parliament.
>
> I am for free trade in religion … With regard to your glorious old city, as I consider six and thirty churches are a far greater number than you can possibly require, I propose that St Peter's, being conveniently situated for the purpose should be immediately converted into a magnificent establishment for printing and publishing the Norfolk News, St Mary's will form a very spacious Baptist chapel and St Clement's will be handed over to the Independents. I would willingly have left untouched the beautiful structure of the Cathedral with its accompanying palace and precincts but I fear they may be required for a Grand Central Terminus for the United Norfolk, Ipswich, Yarmouth, Lowestoft, Dereham, Wells, Fakenham, North Walsham and Aylsham Railways, which we have it in contemplation to construct.[19]

There were three candidates for two seats; Peto, The Marquis of Douro, standing as a Conservative and the Radical J. H. Parry, who had strong Chartist sympathies and presented a significant challenge to Peto. In the end however, he won convincingly with 2,448 votes, to Douro's 1747 and Parry's 1572.

FOUR

Post-Mania Revivals
1848–1852

THE BIRMINGHAM & OXFORD JUNCTION RAILWAY

WHEN the Railway Mania came to an abrupt, and, for many speculators, disastrous end in the autumn of 1847, Peto not only had still to complete his Great Northern and East Lincolnshire contracts but had a few months earlier agreed to undertake another major project. This was the Birmingham & Oxford Junction Railway (B&OJR), a thirty-five mile long broad gauge line, running from Fenny Compton, just north of Banbury, through Leamington and Solihull to Birmingham. The new contract was for the entire line, but with the exception of the tunnel on the approach to the Snow Hill terminus in Birmingham, which, perhaps because of his recent experience at Southampton, he had allowed to be let to another contractor, Branson & Gwyther.

The B&OJR was an early casualty of the financial crash and within a few weeks of work commencing, the company found itself unable to meet the payments due to Peto, the directors deciding in November 1847 to suspend construction completely. Although he went through the motions of threatening litigation, Peto soon realised that there were only two alternatives open to him, either to abandon work, or to take measures personally to rescue the company. Realising that persevering with the new contract would enable him to maintain the momentum of his operations in the forseeable future, and perhaps hoping the depression would be short lived, he decided on the latter course. This would allow him to utilise the plant already in place and avoid disbanding a significant portion of his workforce. But the arrangement he made with the B&OJR directors early the following year, committed him more substantially to the project from the start than any he had previously assisted.

The support took a new form for Peto. The B&OJR would advance him £50,000 in debentures, paying him interest on these for eighteen months, which he would redeem through the sums due to him for the work as he carried it out. There was a clause in the agreement, however, which obliged him to pay the company 5% interest on any of the debentures that had not been redeemed within eighteen months, in effect a penalty should he fail to make the progress expected on the construction of the line. To prevent him proceeding faster than desired, the B&OJR also stipulated that expenditure should not exceed £5,000 per month. The overall result was to compel Peto to execute the work at a considerably slower rate than he had been accustomed in the Mania years; and either to hold a substantial quantity of debentures, or dispose of them, on what still might be a very depressed market.

This deal with the company enabled Peto to make moderate progress on the construction of the line during the depth of the depression, the B&OJR directors duly noting in the minutes in July 1850 that the 'advance' they had made to him two years earlier had been 'absorbed on the works'. The company was then in a somewhat stronger state thanks to the support it was now receiving from the Great Western Railway (GWR) This company had been inconsistent in

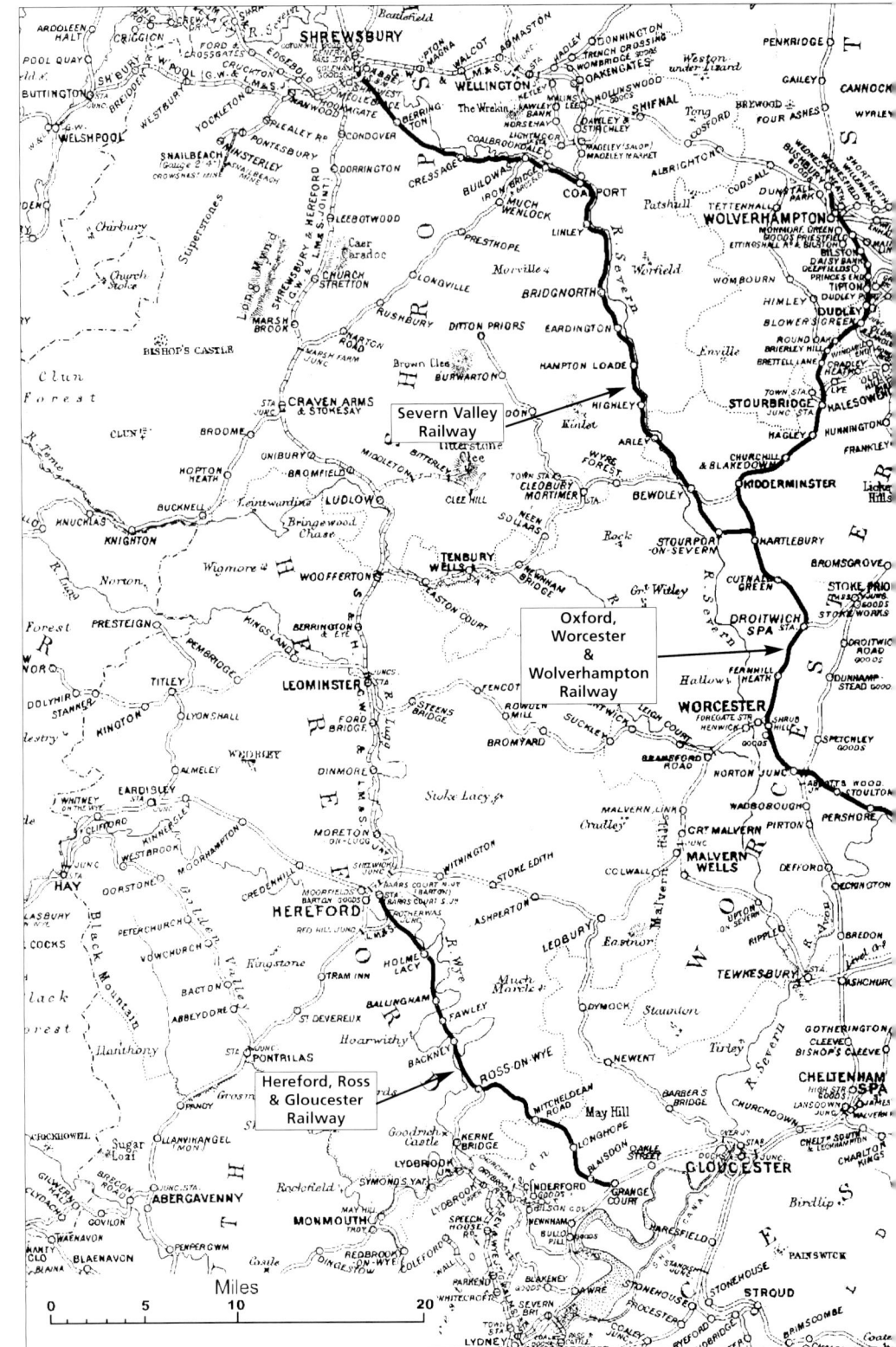

Severn Valley
Railway

Oxford,
Worcester
&
Wolverhampton
Railway

Hereford, Ross
& Gloucester
Railway

Miles

0   5   10   20

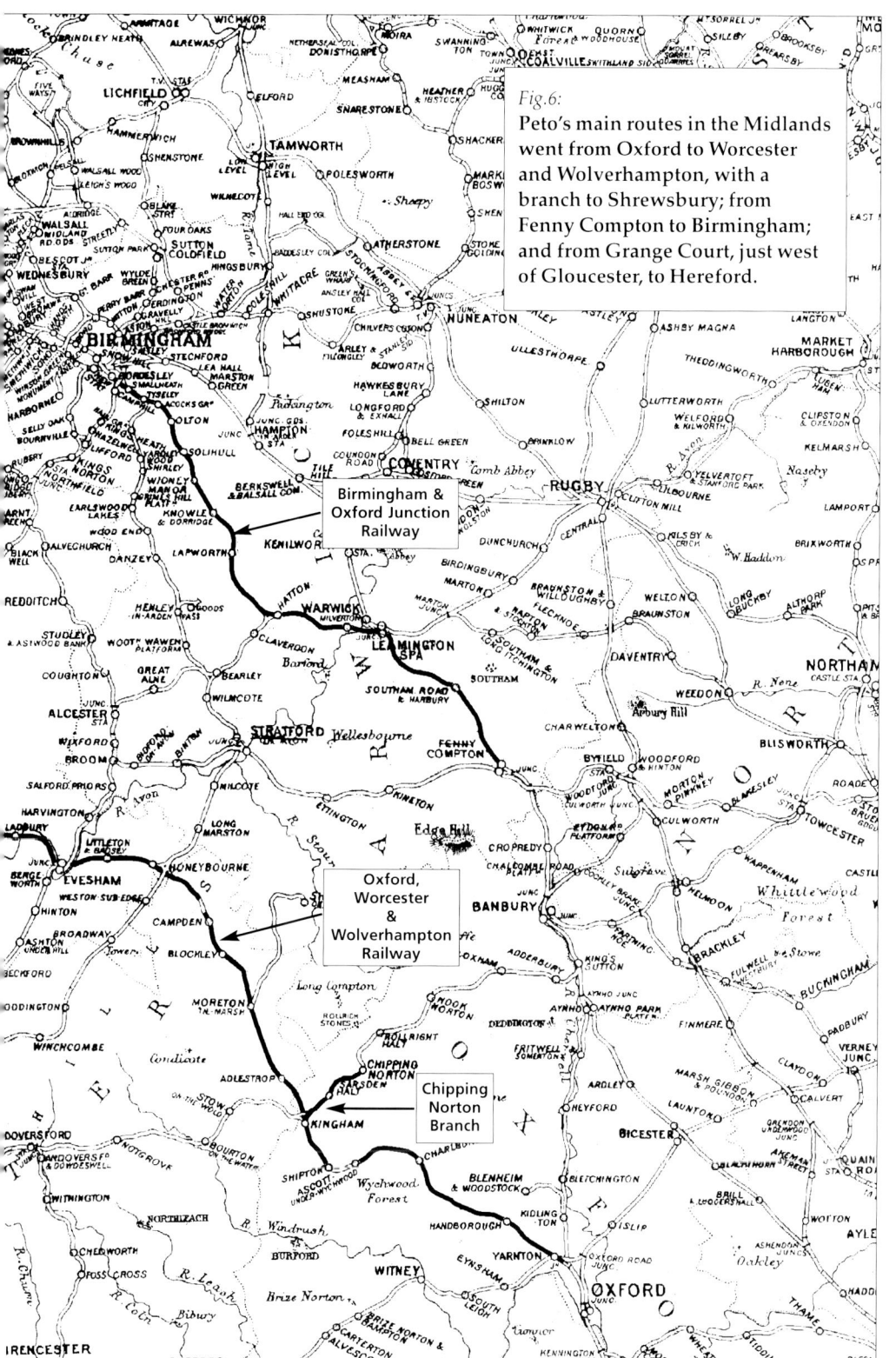

Fig.6:
Peto's main routes in the Midlands went from Oxford to Worcester and Wolverhampton, with a branch to Shrewsbury; from Fenny Compton to Birmingham; and from Grange Court, just west of Gloucester, to Hereford.

Birmingham & Oxford Junction Railway

Oxford, Worcester & Wolverhampton Railway

Chipping Norton Branch

its backing of the project, having initially supported it, then changed its mind, making Peto's intervention necessary, before finally grasping at the opportunity it provided for the broad gauge to reach the heart of the West Midlands. Consequently Peto's next 'advance', in addition to the company's own mortgage bonds, included an allocation of GWR stock, which was more easily disposed of than if it had been in a company whose line was still under construction

The B&OJR did not prove a particularly easy line to construct. At its southern end the route crossed the jurassic limestone ridge running from Warwickshire into Northamptonshire. The engineer, Brunel, in his original plans proposed to tunnel through part of this at Harbury but, after several collapses, was obliged to replace most of this with a cutting, which necessitated the removal of about 1½ million cubic yards of earth. According to Peto, this was the largest work of its kind in the world at the time. Other significant works were a bridge 160 feet long across the river Avon, an aqueduct to carry the Warwick and Napton Navigation over the railway and considerable lengths of brick viaduct at Leamington and Bordesley. These difficulties and the restriction of expenditure resulted in the the line taking more than five years to complete, not being finally opened until the autumn of 1852.

In the end Peto received more than £800,000 from the company, making the B&OJR, at over £20,000 per mile, the most expensive on a mileage basis of all the lines he had so far built, with the exception of the Blackwall, which had been a special case being on viaducts for almost its entire length. Whilst some of the high cost of the B&OJR can certainly be attributed to the engineering involved, including Brunel's decision to lay additional rails to accommodate both broad and standard gauge trains, it also seems likely that the company paid quite dearly for the financial assistance it received from Peto, due in part at least to the works being executed entirely, or at least partially, on a schedule of prices. How profitable the contract really proved is nevertheless uncertain, as much would have depended on what terms he disposed of the

debentures and other stock, which constituted such large proportion of his payments. When it came to his own costs it would have been to Peto's advantage that most of the construction was carried out in a depression, especially if there was no adjustment to the rates he could charge for labour and materials after the contract was re-negotiated in 1848. If he did benefit in this way, and there is nothing in the records to suggest either way, the B&OJR directors did not complain about his charges. Nor was Peto, as far as is known, allowed to indulge to any appreciable extent in additional works, which may indicate closer supervision of the work than with the the the Great Northern and Eastern Counties works.

THE CHESTER & HOLYHEAD AND MOLD RAILWAYS

PETO had carried out the B&OJR contract, like the GNR, in collaboration with his brother-in-law, Edward Ladd Betts, although they did not enter into formal partnership until after Betts had completed his work for the Chester & Holyhead Railway (C&HR) in 1848. This had included the first eight miles of the line out of Chester; as an adjunct to which Betts had undertaken to construct a branch to the Flintshire market town of Mold, for a nominally independent company, which also included a short branch to serve some quarries at Ffrith. By the summer of 1848 however, the Mold Railway (MLR) was running short of funds and faced with the imminent prospect of having to suspend operations. This would almost certainly have happened but for Peto's timely intervention, with the offer to subscribe £180,000 of the company's shares. This was considerably more than Betts's original contract price of £146,0000 and a substantial speculation on Peto's part.

The immediate consequence was to enable Betts to proceed with the work in hand and also to build a number of additional short branches serving some colleries in the neighbourhood, these additional works raising the total cost of the undertaking to £155,000. The railway was opened within a year of Peto's involvement, but by 1850 had been absorbed by the C&HR, which *de facto* made him a substantial shareholder in the larger company. The C&HR was not, however, in a financially healthy state either, having expended very large sums on its construction and not as yet able to earn sufficient revenue to service the capital outlay. The directors consequently welcomed Peto's arrival on the scene, with his apparent ability to provide funds for the ailing project, and proceeded to elect him their Chairman in 1851. Although this post, and the one he also obtained that year with the Norfolk Railway, elevated his status in the railway hierarchy, it brought problems he was not accustomed to face as a contractor.

Besides the financial crisis which dogged the C&HR throughout the eight-year period he was at the helm, Peto was also involved in prolonged negotiations with the Government over the arrangements for the conveyance of mails from Holyhead to Ireland, and with the London & North Western Railway over a possible amalgamation of the two companies. When terms for this were eventually agreed in 1859, he was almost certainly relieved to be able to retire from a post he appears not to have particularly enjoyed and which did not enhance his reputation. Nor, as far as can be known, did it bring him any significant personal gain. He would seek no further chairmanships of railway companies, preferring henceforth to exert his influence behind the scenes, where he was not directly accountable to directors or shareholders.

Whilst Chairman of the C&HR, Peto had not been able to resist the opportunity to speculate personally in two projects that aimed at the same time to bring additional passenger traffic onto the railway. Both sought to exploit the scenic attractions of the area it served. He engaged the services of his friend Joseph Paxton to design an extravagant 300-bed hotel to be placed at a vantage point overlooking the Menai Straits, close to Robert Stephenson's majestic Britannia bridge which carried the C&HR over to Anglesey. The appropriately named Britannia Hotel was to be built on land surplus to the needs of the railway and be connected with it by a short branch line. Although the foundations were laid, insufficient funds were forthcoming to proceed with what would have been an even more flamboyant gesture than his earlier speculation at Colchester. Considering the fate of that enterprise, he was probably wise not to have proceeded with this venture. Even less progress was made with another hotel planned to be built at Rhyl. In the end the only private project he completed was a modest row of terraced houses facing the seafront and adjoining the railway at Abergele, which were designed to be let to summer visitors. Peto no doubt hoped that these would become the basis for a new resort served by the C&HR. It was to be overshadowed however, by the growth of neighbouring Llandudno, the sad state of the original properties today bearing testimony of another unfulfilled Peto dream.[20]

THE OXFORD, WORCESTER & WOLVERHAMPTON RAILWAY

As the 'fifties dawned and the depression in railway construction continued, Peto had cause to ponder on whether parting with Grissell had been an entirely wise move. There was some light on the horizon however, as the investment market was at last showing signs of recovery, which encouraged at least a few railway companies to consider resurrecting schemes they had been obliged to abandon earlier. One such company was the Oxford, Worcester & Wolverhampton Railway (OW&WR), which had begun work during the Mania on a broad-gauge trunk line from Wolvercote, just north of Oxford, through the Cotswolds and the Vale of Evesham to join the Birmingham & Gloucester Railway (B&GR) near Worcester, and proceed from there to Wolverhampton by way of Droitwich and Tipton. Despite expending more than £3 million, all the company had to show for its efforts when work was halted in 1847, was a link from the B&GR to Worcester, although some sections of track elsewhere were partially finished.

The OW&WR works had earlier been let in fairly short sections to a number of different firms, but when the project was resuscitated late in 1850, the company decided to pay off most of the original contractors and engage Peto & Betts and another major firm, Treadwell Brothers, to complete all the outstanding work, with the exception of two of the Cotswold contracts. These were a section of track near Shipston-upon-Stour and the tunnel at Mickleton near Chipping Campden, where the original contractors were re-engaged. Peto took over the remainder of the work between Wolvercote and Worcester, as well as the line from Tipton to Wolverhampton, whilst Treadwells agreed to complete what remained unfinished between Worcester and Tipton. Both new contractors were however required to invest in the company. Peto agreed to subscribe £20,000 of shares, on condition that the 'calls' on these were met out of the payments due to him for the work as he carried it out. This appeared a preferable arrangement from his point of view to that currently applying to the neighbouring B&OJR, as his overall commitment would be less and the proportion of the cost of the works to be paid for in cash greater. Also, there were no restrictions on the speed with which he could proceed; on the contrary he was obliged to complete his part of the Wolvercote–Worcester line within eighteen months of starting work and the Tipton to Wolverhampton line within twelve months of finishing the former.

His support for the company had ensured that Peto did not have to compete with other firms for the contract, although he was obliged to undertake all the work for a fixed sum. The works between Wolvercote and Worcester were to be executed for £395,000 and the Tipton–Wolverhampton line for £229,000, the prices in both cases including the stations. This was a sensible precaution on the company's part in the light of the experience of some of their predecessors. There was also a clause in Peto's contract limiting the outlay on rails to 10,000

tons per mile. The only concession granted him was not to be liable for the cost of any damage arising from subsidence between Tipton and Wolverhampton, a wise provision on his part, as this section of the line crossed a coal mining district.

The apportionment of the works had one undesirable aspect as far as Peto was concerned. Although he would be constructing in all eighty miles of track, this was in four distinct parts, each isolated by undertakings in the hands of other firms, the surviving Shipston contract separating his Wolvercote works from those on to Mickleton, with the tunnel there an additional break in the remainder of the line to Worcester. Treadwell's contract lay between this and his from Tipton to Wolverhampton but was no problem, as Treadwells, like him, had a deadline for completing their undertaking. Meeting the company's requirements between Wolvercote and Worcester would be easier for Peto, if the Shipston and Mickleton contractors completed their tasks early enough for him to co-ordinate his own operations. The Shipston contractor had comparatively little left to do and readily complied, which was more than could be said of his opposite number at Mickleton. The works there were in the hands of R. M. Marchant, whose firm had replaced another just prior to the cessation of work in 1847, and had consequently been put to considerable expense for comparatively little return. Marchant consequently considered himself entitled to considerably more compensation than the OW&WR was prepared to pay when it revived the project. Having his plant in place put him in a strong bargaining position but the company, obliged to keep expenditure to a minimum, refused to give way. Marchant's response to this treatment was to slow down work on the tunnel.

With Marchant's inactivity threatening the prospect of the railway being completed in the time specified, which was crucially important for the company, as it wanted to begin earning revenue as soon as possible, the OW&WR directors instructed their engineer, Brunel, to take the necessary measures to evict Marchant. Matters came to a head over the weekend of 20–23 July 1851, Brunel arriving at the mouth of the tunnel on the Friday with several hundred navvies. This threat of force did not have the desired effect however, Marchant remaining in the tunnel with a sizable body of his own men. Warned that there was serious danger of a violence clash, the local magistrates despatched one of their number to the scene, who duly read the Riot Act. This may have had the desired effect, as a period of stalemate ensued, with the opposing forces biding their time in the mid-summer heat. By the Monday morning, however, the situation had changed dramatically, Brunel's men having by this time been substantially re-inforced by navvies marched across country from Peto's works on the neighbouring B&OJR line. Outnumbered by ten to one, Marchant capitulated peacefully, Peto's intervention having secured a near bloodless victory for Brunel and the company. Marchant for his part held Peto more to blame for his downfall than either the ow&wR or Brunel, giving vent to his feeings in a letter to *The Railway Times* :

> I may leave Messrs Peto & Betts to defend themselves against the charge of having consented to the march of over two thousand men on a Sunday for the purpose of taking possession of my property by force.[21]

Marchant was endeavouring to expose Peto as a hypocrite for consenting to the employment of labour on a Sunday, something most of his fellow Nonconformists were strongly opposed to. Peto for his part appears to have had few qualms in separating the needs of business from any religious commitment. The indictment of Peto for his actions that weekend was to have connived in the eviction of a fellow contractor, who had a justifiable case against the company. But with a significant investment in the ow&wR, there could be little doubt from the start where his loyalty lay. He wanted Marchant's removal to make his and the company's tasks easier to accomplish. This was almost certainly his sole aim, as he was quite happy for the company to complete the tunnel works itself. If Peto had no feelings of conscience over the Mickleton affair, Brunel possibly did, as he resigned as engineer

shortly after. He was replaced, perhaps not surprisingly, by John Fowler, with whom Peto had worked three years earlier on the East Lincolnshire Railway.[22]

Peto experienced some difficulties in constructing his portion of the ow&wR. During the winter of 1852/3 there was severe flooding along the Avon valley, some of the embankments and a number of the bridges being washed away. He had to make good this damage, which involved him in considerable expense, the viaduct at Addington, west of Evesham, having to be completely rebuilt. Also, after the line had been completed between Oxford and Worcester, the Board of Trade Inspector declared a section of the permanent way west of Wolvercote, where mixed-gauge track had been laid, to be unsafe. This not only necessitated additional expense but also delayed the opening of the line for several weeks. Peto was not penalised however, for failing to meet the deadline set in his contract; either because Fowler was more at fault, or, as seems more likely, the company had by this time come to depend too much upon his financial assistance. Both Peto and Treadwells had each been obliged to subscribe a further £25,000, this time in preferential shares. Peto had in addition assumed a more important role in the company's affairs, arranging and guaranteeing personally a number of loans for it in the City, involving amongst others the Union Bank and the discount house Overend & Gurney. This is the first recorded instance of his involvement with these financial institutions, which would figure prominently in his later financial activities.[23]

The railway was finally opened from Oxford to Worcester in July 1853 and through to Wolverhampton a year later. Now freed of the constraints imposed on him by the contracts, Peto first obtained a place on the ow&wR Board and in 1856 was elected Deputy Chairman. One of his reasons for seeking office was almost certainly to encourage his fellow directors to embark upon two major extensions of the railway, which, had they come to fruition, could have provided him with lucrative new contracts. One was the Cheltenham Direct Railway, which besides providing the spa town with a shorter route to London than the existing line through Swindon, would have enabled the OW&WR to tap the lucrative South Wales coal traffic. The project failed to materialise however, due in large part to the opposition of a number of influential landowners. The other scheme was even more ambitious; an extension from Wolvercote eastwards to link the OW&WR with the London & South Western Railway (L&SWR) at Brentford, which would have competed with the GWR for Oxford–London traffic. This scheme failed to obtain Parliamentary approval. In the end the best Peto could manage in the way of new lines was the contract for a short branch from Kingham to Chipping Norton, where he was obliged to subscribe £16,000 out of the £26,000 share issue.

Peto remained Deputy Chairman of the OW&WR up to the time of its absorption into the larger West Midland system in 1860. Although the OW&WR did quite well out of this, as so often, it is not possible to know to what extent Peto benefited personally, although it seems likely that he retained a significant stake in the company to the end.

THE HEREFORD, ROSS & GLOUCESTER RAILWAY

Not all the projects suspended at the end of the Mania were as easily revived as the OW&WR, the time allowed for construction in their Acts of Parliament having in some cases expired before the necessary funds became available. The Hereford, Ross & Gloucester Railway (HR&GR) was in this position when the directors approached Peto for assistance to relaunch the scheme. Although this had originated in the Mania years, virtually no work had been done before the economic tide turned, so a fresh start could be made when a new Act was obtained in 1851. Brunel, who was also engineer for this broad-gauge project, took advantage of the situation to make some alterations to his earlier plans with a view to reducing the cost of construction, the most important being to route the line through

Ross-on-Wye rather than Monmouth. Despite these economies, the company had still been unable to raise the funds it needed on the open market and consequently made its approach to Peto. After protracted negotiations he agreed to invest substantially in the company, and also agreed to build the line partially on credit.

The terms of the contract were finally agreed in May 1852. As well as his partner Betts, Peto undertook the HR&GR in collaboration with Thomas Brassey. Having the backing of the two greatest railway contractors of the day was invaluable to a small struggling company like the HR&GR, whilst Peto would have welcomed the additional resources Brassey could bring. There was an element of self interest behind Brassey's agreement to join, as he had, as part of his arrangement with the Shrewsbury & Hereford Railway (S&HR), agreed to lease that line from the company when he had completed it. Any additional traffic coming on to the S&HR from the HR&GR, would obviously benefit the lessee. Although Brassey agreed to back the new scheme financially, he seems to have been happy to leave its construction to Peto & Betts.

The contractors agreed to subscribe more than half of the HR&GR capital, £137,000 out of the £275,000 provided for in the new Act, whilst the company would pay them half the costs of the works in debentures, a further third as credits to set against the 'calls' on their shares, leaving only a sixth to be paid in cash. There was a clause in the contract, however, which limited their overall liability on the 'calls' to £103,000, with not more than £15 on each £100 share to be due at one time. Overall the liabilities assumed by the contractors were even greater than Peto had accepted five years earlier with the Mold company, which was probably why he was pleased to have Brassey's support.

Unusually Peto did not move quickly to begin work and appears to have deliberately drawn out the negotiations over the terms of the contract, which caused the HR&GR directors to accuse the contractors of 'dragging their feet'. The reason for the delay was soon revealed, Peto telling them that, although he was 'anxious to proceed with the works', he wished to put one of his 'best men, Mr Watson, who has been and is still conducting works of the Birmingham & Oxford Junction Railway now nearly completed' in charge. He might have added that it would also be more economical to transfer plant and operatives from a line he had just finished, than start work from scratch on the new project. With the company so dependent upon the contractors, the directors could only wait upon Peto's convenience

The start of work was further delayed by disputes with some landowners but once these were resolved, Peto pressed ahead with his customary vigour. The line was the most difficult he had undertaken up to that time, as it cut across the grain of the country, necessitating deep cuttings and high embankments, as well as four bridges across the river Wye. There were also four tunnels, at Lea 771 yards, Fawley 540 yards, Billingham 1,210 yards and Dinedor 110 yards. Lea Tunnel proved the most troublesome, work here being held up for several weeks due to flooding from underground springs. Brunel had designed the viaducts as wooden trestle structures, which made them fairly easy to construct and cheaper than wrought iron; but as Captain Tyler, the Board of Trade Inspector, subsequently remarked, they were too flimsy to prevent excessive vibration when trains passed over. The Inspector did not comment adversely however, on the way these had been constructed or on any other aspects of the contractor's work.

Although the short section between Grange Court, where the HR&GR left the South Wales Railway, to Lea Tunnel, was opened for traffic in 1853, the rest of the line took a further two years to complete. The final opening through to Hereford was delayed further by the need to obtain the agreement of the Shrewsbury company to the layout of the new joint station at Hereford. Until this was completed a temporary building, erected by Mr Eassie, a Gloucester contractor who specialised in such work, had to suffice. It appears that the company chose the cheapest option, although Peto's contract had excluded the stations anyway.

The HR&GR remained chronically short of funds throughout the construction period, which made it difficult to meet even the comparatively small cash element in the contractors' monthly certificates, and by its completion there was a deficit of £33,000 in this account. It was agreed in the end that the deficit would be paid off over a ten-year period with 5% interest. The contractors had meanwhile come to an arrangement amongst themselves, whereby Betts and Brassey consolidated their investments in the company and transferred this to Peto, making him by far the largest shareholder in the company and in a position to determine the actions of the Board. Although he preferred to exert his influence behind the scenes, he nevertheless took the precaution of having Betts made a director, so that he had a proxy present at Board meetings. Betts had fulfilled a somewhat similar function earlier, it will be recalled, with the Eastern Counties Railway.

With such a large stake in the HR&GR, its continuing financial difficulties must have caused Peto concern, not least because the shares he had accepted were both depressed and not easily disposed of. He was also obliged to raise a loan on the company's behalf in 1857, needed to enable it to repay the outstanding debentures, a large number of which were as likely as not to have been in his name. Peto had probably hoped that by this time the HR&GR would have been absorbed by a larger company – which may well have been what had motivated him when he agreed to accept his co-contractors' shares – but he had to wait until 1862, when the GWR finally agreed to take over the company. Again it is difficult to know how well Peto fared personally as a result, but any eventual profit from his shares would have had to be set against the cost of supporting the company for more than a decade. The HR&GR for its part could not blame Peto for its difficulties; he had constructed a difficult line at a modest price – just over £10,000 per mile, no more than his much easier East Anglian works, although it must be admitted that these had been built in the expensive Mania years.

FINANCIAL SITUATION

By employing his own reserves of capital and creditworthiness, Peto had managed to ride out the the post-Mania depression and keep his contracting force largely intact. The first question which must be asked, was whether he could really afford to accept the consequent liabilities. These amounted in all to more than £300,000, although not all at the same time, and in the case of the HR&GR, he had the not inconsiderable backing of Brassey. On the other hand the dissolution of the partnership with Grissell had deprived him of an important alternative source of income, which was almost certainly not compensated for by the resources Betts brought to the new partnership. As it happened, at this juncture in his career Peto was questioned for the first time as to the real extent of his personal wealth, although the disclosure was in no way forced upon him by his business activities. In 1850, partly as a result of his friendship with Henry Cole, he agreed to support the Great Exhibition planned for the following year, heading the list of guarantors to the tune of £50,000, and was subsequently made one of its financial commissioners by Prince Albert. Those in charge of the project felt obliged, however, to make some discreet enquiries as to his ability to meet such a large liability and approached his solicitor, Mr Lawrence for reassurance on this matter; to receive the reply that his client was worth between £600,000 and £700,000. Although this may have included the shares he held in various railway companies at the time, Peto could, on this evidence, have afforded to act as he did during the post Mania period, even if at times he had been dipping more deeply into his reserves than was entirely prudent.

Altogether Peto with his partners, had invested about £300,000 in five companies between 1848 and 1852, for works worth in the region of £1million. Assuming they made the 10% profit he regarded as the minimum acceptable, they would have netted a return of about £100,000, although on a limited proportion of this in cash. He would also have hoped to obtain a significant return from the shares he came to hold in the companies for which he had worked. Besides sealing the deals on these contracts, the investments must have also attracted him as opportunities for speculation in railway stock and a fulfilment of the desire he had expressed so unequivocally at the time of his break with Grissell, to become a fully-fledged man of capital. It had also enabled him to extend his activities into railway management, a distinct step up in the railway hierarchy for most contractors, although not so much so for a major figure like himself. The test ahead was whether he would be equally successful in his new roles as he had been in the earlier; on his own testimony all the early contracts, with the exception of Hanwell, having been profitable. Whilst he discovered at quite an early stage that company management was not his particular forte, the verdict on his ability as a financier was still open and would not be resolved for more than another decade.

Although Peto had not made any concerted effort to diversify his business interests since the Mania, which would have been one way of safeguarding himself against unforseen financial difficulties arising from the railway undertakings, he had nevertheless extended his activities beyond the contracts themselves. Besides the hotel and property speculations already mentioned and the ill-fated North of Europe Steam Navigation Company, by the early 'fifties he had acquired a major stake in one of the country's major industrial complexes. The Clay Cross Company was in every way a railway creation; the deposits of coal and iron ore in this part of north Derbyshire had been discovered by Robert Stephenson and his father, whilst they were constructing the North Midland Railway. The mines, quarries and blast furnaces developed on the site, were largely financed by entrepreneurs with railway interests, including George Hudson and William Jackson, Peto's predecessor as Chairman of the C&HR. Peto invested £20,000 and in due course became a director and then Chairman of the company. Unlike the more speculative ventures, it proved a wise investment, the shares he held having more than trebled in value when he sold them ten years later.

POLITICS

PARALLEL with his business interests, Peto had been actively pursuing his political career, the highlight being his seconding of the address in reply to the Queen's Speech in 1851, when he made a point of extolling the virtues of the new mechanical age and stressed the civilising influence of steam, which he said was 'uniting the distant portions of Europe' and in his view making 'states more desirous of peace.' During the 1847–52 Parliament he embarked upon a number of crusades he would pursue throughout his time in Parliament. One was the liberalisation of trade, making a strong attack upon the tax on bricks, which he said mitigated against improvements in working-class housing, although he failed to add that its abolition would also benefit him as a contractor. Another was the need to improve the sanitation of London, which he said should be financed by a £1million loan from the Treasury. These were issues which already attracted mainstream political support but he found himself more isolated when he attacked the Admiralty for failing to embrace the latest technology, and especially for continuing to construct wooden warships. He also sought unsuccessfully to legalise the right of dissenters to be buried in parish churchyards. Not forgetting that he was very much part of the railway interest in the House, he supported Joseph Locke's Bill to allow more trains to run on Sundays and advocated the development of railways in India. The narrower concerns of his own Chester & Holyhead company were undoubtedly the main reason why he strongly backed a motion to improve postal communications with Ireland.

Peto sought re-election at Norwich in the 1852 general election, when he ran with fellow Liberal Edward Warner against two Conservatives, the sitting member, the Marquis of Douro, and Colonel Dickson. This time economic rather than sectarian issues dominated the campaign, which benefited Peto, as his free trade views assured him of the support of many who had voted Radical in 1847. He was not however prepared to advocate universal franchise, which his Radical opponent had campaigned for at the previous election. He sought a popular mandate by accusing his opponents of advocating dearer food, pointing out that owning a country estate, did not prevent him from embracing the cause of tariff reform. In reply the Conservatives endeavoured to revive the Janus image they had portrayed of him at the previous election, one of their bills proclaiming:

SIG SMOOTH = FACE PETRO
The great telegraphic Necromancer will delude the public. In his performance he will put a large loaf marked Free Trade Norwich into his hat and any person … will be at liberty to take up the hat when they will find a much smaller loaf with Protection Somerleyton on it.[24]

It had been an empty gesture on Peto's part anyway, as he did not depend upon his Somerleyton rents.

Although declaring that universal male suffrage was not yet practical, Peto did however support the extension of the franchise to those in towns who paid over £10 a year in rent. He also advocated a secret ballot for parliamentary elections, which he had not been prepared to do five years before. The Liberals took a particularly strong line against bribery at this election, their agents making a door-to-door canvass of the electors trying to persuade them not to accept bribes. This did not prevent Peto from laying on refreshments, albeit of the non-alcoholic variety, at the two election meetings described in the press as soirées, which were held at the West End Retreat. With characteristic flair, he arranged for a pre-fabricated pavilion that had been used the previous year at the Great Exhibition, to be erected in the grounds, but unfortunately the glass roof made it so hot inside that the pro-ceedings had to be reconvened in the open air! He need not have gone to so much trouble, as he and his running mate were easily elected, with 2186 and 2134 votes respectively, to 1588 and 1454 for their opponents.

FIVE

Railway Promotions
1852–1859

THE LONDON, TILBURY & SOUTHEND RAILWAY

PETO combined forces with Brassey again in 1852, this time to promote a new railway company rather than resurrect a lapsed project. From the outset the London, Tilbury & Southend Railway (LT&SR) was a speculative venture, as it would cross the sparsely populated estuarine marshes of Essex. Tilbury at the time was little more than an historic fort commanding the entrance to the Thames, whilst Southend, although already an established watering place served by river steamers, was still a comparatively small town. Peto's friend Bidder had already realised that the port of London would have to extend eastwards in order to accommodate larger vessels, and ten years earlier had promoted the Eastern Counties & Thames Junction Railway to Thames Wharf. Despite the warnings of the sceptics who dubbed the line 'Bidder's folly', the project had been a success, probably giving Peto the idea he could develop the concept further. But with the investment market still in the doldrums and his own resources stretched, some way had to be found to induce the public to invest in the scheme. The solution he and Brassey arrived at, was to guarantee the investors a fixed return for an extended period, by agreeing to lease the LT&SR from the company before they began building it. As already mentioned, Brassey had entered into a similar arrangement with the Shrewsbury & Hereford Railway (S&HR), in that case agreeing to pay the shareholders 4½% interest and half the profits for eight years.

The LT&SR shareholders obtained more generous terms from the Peto/Betts/Brassey consortium than their S&HR counterparts had from Brassey, being guaranteed 6% interest for twenty-five years, which included the time when the railway was under construction. As a result the contractors were compelled to pay out significant sums before they received any income from the line; and there was no certainty, even when it was up and running, that the revenue would be sufficient to pay what was at the time an above average dividend. In addition the shareholders were also entitled to half the profits of the railway. Looked at from the lessees' point of view, the LT&SR was a twenty-five year liability, which had to be met from the profits on the contract and the revenue from the railway once it was completed. There was always the hope, however, that the LT&SR might be sold on advantageous terms long before the lease expired.

From its conception the LT&SR had the support of the neighbouring London & Blackwall and Eastern Counties companies, whose directors made up the Joint Committee set up to manage the railway. Individual investors in the scheme were not directly represented on the Joint Committee, having to be content with the right to attend the periodical shareholders meetings. Although the contractors did not have a place on the Joint Committee either, they were guaranteed access to all its deliberations and Peto made certain he was consulted from the start on all financial matters. He also appears to have been largely responsible for the execution of the works, as his agent Mr T. White, who had earlier worked on the Norfolk and Great Northern lines, was put in charge.

Under the terms of the contract agreed with the Joint Committee in July 1852, the forty-one miles of track running from the Forest Hill Junction with the ECR just east of Stratford, to Southend via Tilbury, were to be constructed for £400,000. This conformed to the £10,000 per mile initial price that Peto appears to have accepted as a norm by this time. It included the cost of land as well as parliamentary and legal expenses. But Peto soon realised that he had been too generous to the company and within a month, wrote to the Joint Committee, telling it that the contractors would only be able to earn a 4% profit, when they expected a minimum of 5%; and asking for the contract price to be increased by £25,000. He did not mention the liabilities they had accepted under the leasing arrangement but he may not have considered that diplomatic at the time, in view of the company's hope of its share in the profits. He did however conclude with the threat, should the directors not accede to his request:

> With pressure upon us for works and
> a rapidly rising market, especially in labour,
> we are only kept from declining altogether,
> by our sense of what is due to the Committee
> from the honourable understanding which
> has existed, and which we desire to carry
> out to the greatest extent.[25]

After initially protesting, directors complied, probably realising that overall they had not obtained a bad deal, although the contract had not specified anything about the cost of stations, or other additional works.

Having got his way over the price of the works, Peto did not relax his hold over the LT&SR. He made certain from the start that he was consulted by the Joint Committee over share issues, including the timing of the 'calls', which was a justifiable precaution on his part, as the contractors were responsible for the interest. Before long however, his involvement in the finances of the company took on a more dubious aspect. From an early stage in their dealings he persuaded the Joint Committee to advance him cash and debentures in excess of what was due to him from the works; offering as security either LT&SR shares he had already subscribed,

or stock of companies he had worked for previously. In January 1853 the LT&SR advanced him £70,000 on the collateral of 13,593 LT&SR shares and 6,875 in the Hereford Ross & Gloucester Railway (HR&GR). In 1855 he obtained a further £45,000 in return for more shares in the HR&GR and some Eastern American Railway bonds; the same year he also used 2,000 of his shares in the Grand Trunk of Railway of Canada for the same purpose, when the desperate financial plight of that company made them a very questionable security. In addition to using the LT&SR as his private bank in this way, Peto contrived at the same time to be paid in advance for much of the work he carried out, substantial cash sums being involved, £35,583 for instance in February 1855.

The railway was opened from Forest Gate Junction to Tilbury in 1854 and on to Southend the following year, after some problems had been experienced in crossing the estuarine mud. By this time Peto had embarked upon a major additional work, a 'cut-off' line from Barking to Gas Factory Junction, which was designed to give the LT&SR direct access to the Blackwall company's new City terminus at Fenchurch Street. This was expensive, as it involved considerable lengths of brick viaduct, the contract price being about £200,000, which, like the rest of the additional works, was covered by the terms of the original lease. These included doubling the track between Tilbury and Southend, a branch to serve some wharves at Thames Haven and a warehouse at the Minories, which alone cost £30,000. This brought the total expenditure on the railway to over £700,000, which almost certainly considerably enhanced the contractors' profits but at the same time increased their long-term liabilities. Also, once the railway was finished, Peto could no longer rely upon the company to assist him with his own finances, although the Joint Committee was prepared as late as July 1856, to provide the collateral he needed for a loan from the London and Westminster Bank.

Rather than operate the LT&SR themselves – which neither Peto nor Brassey had the facilities or expertise to do on their own – the lessees came to an arrangement with the ECR, which was well-placed to provide the necessary services as an adjunct to its own line. The traffic was light from the start and failed to provide enough revenue to cover the interest charges, making the lease a long-term liability for the contractors. Consequently there were no surplus profits to share with the proprietors. Not wishing to depress the value of the shares nor antagonise the shareholders, Peto at first declined to publish the traffic returns, telling the Joint Committee, that these did not provide a true picture of the company's prospects. When the shareholders pressed him further, he tried to buy them off with an extra ½% interest on their shares. Unwisely from their point of view, they rejected this offer out of hand.

Peto then made a bold move to increase the number of passengers using the railway. Where a short while before he had provided accommodation for visitors at Lowestoft to boost traffic on his ailing Norfolk line, he now tried to persuade Londoners to forsake the increasing noise and grime of the capital for the clean bracing air of the Essex coast, and to commute daily along the LT&SR. Although many of the better-off were already quitting the City and living further from their place of work, long-distance commuting was comparatively rare in the middle 'fifties. This did not deter Peto, who saw the potential advantages of being one step ahead of the market, and along with Brassey, acquired some land adjacent to the station at Southend, upon which they proceeded to lay out the Cliff Town Estate. This was designed to meet the requirements of middle class families, boasting amongst its attractions the latest sanitary arrangements. In an attempt to beautify the surroundings, trees were planted, an esplanade built and the somewhat decrepit pier acquired, along with its horse-drawn tramway. Hand in hand with the building project, the developers provided season tickets at attractive rates and other concessions to encourage residents to patronise their railway. Unfortunately few chose to make the long and rather uncomfortable journey to London in the first two decades of the LT&SR's life. The streets of large, solidly-built town houses that greet the traveller leaving the station at Southend today,

is a reminder of an idea, which was before its time and did very little to relieve the financial plight of the LT&SR and its lessees.

Still burdened with heavy interest payments to the shareholders, the only option remaining for the lessees, was to reduce the costs of operating the railway. This inevitably excited the wrath of passengers, who over the years were obliged to suffer what came to be regarded as one of the worst services in the country. The general run-down of the line also fuelled the existing hostility of the shareholders, who considered the lessees were depreciating the company's assets for their own ends. Nevertheless an offer from Brassey in 1862 to hand back the shares he and Peto still held in the company, if the lease was terminated, was rejected. A proposition made to the ECR and L&BR that these companies should jointly take over the lease for 999 years and pay the proprietors 4½% interest, met with the same response. Earlier in a fit of pique the shareholders had called Peto to task over the expenditure he had incurred in purchasing a vessel for the service that had been instituted between Tilbury and Gravesend. He was obliged 'under protest' to refund £4,000 to the company and also reduce the tolls on this service.

Relations became even more strained with the decision of the shareholders in 1863 to set up a Committee of Investigation to examine working expenses and tolls. This only confirmed the unprofitable state of the railway, without offering any remedy. The LT&SR continued in this unhappy state for the remainder of the 'sixties, with mounting complaints from its users, whilst a request from the lessees to be permitted to raise tolls was turned down by the House of Commons. In 1866 his personal financial circumstances made it necessary for Peto to hand over his stake in the lease to Brassey, for whom it appears to have remained a liability up to his death in 1870.

The Victoria Dock development at the London end of the line may have provided some compensation for the LT&SR's poor performance and the initial failure of the property speculation at Southend. Bidder was the engineering brains behind the scheme, which had its origins in his earlier Thames Wharf enterprise, but was on a much larger scale. The site at Plaistow was excavated to provide facilities for the largest vessels afloat, with the most up-to-date machinery to handle their cargoes and there was also direct rail access to the individual wharves. Part of the development took place on land fronting the Thames owned by the North Woolwich Land Company, in which Peto had taken a stake earlier and it was only when he and his LT&SR partners became actively involved, that the new dock venture really got under way. In 1852 – the year they began the railway – Peto, Betts and Brassey joined the Board of the Victoria Dock Company and agreed to execute the work. This took nearly three years and in the end involved an outlay of nearly £1million. Despite the cost, it proved an immediate commercial success, although it is not certain to what extent Peto and his colleagues benefited. The new facilities were sufficient a threat to the older docks, to induce their owners to buy out the company in 1864.

THE WEST END OF LONDON & CRYSTAL PALACE RAILWAY

THE improvement in the investment market, which had assisted the launching of the LT&SR in 1852, showed signs of developing into a mini railway boom by the following year, producing in its wake a plethora of new schemes. One of these was the West End of London & Crystal Palace Railway (WEL&CPR), which planned to build a line from a terminus on the south side of New Chelsea Bridge to Sydenham, where it had been decided to re-erect the Crystal Palace as a major venue for exhibitions, concerts and other public events. Besides overseeing the relocation of his Hyde Park masterpiece on its new green-field site overlooking the Surrey hills, Joseph Paxton was one of the original promoters of the railway. Paxton had been on friendly terms with Peto for a number of years, and was also indebted to him for commending his original design for the Palace to his fellow commissioners. As a supporter of

the 1851 Exhibition, Peto would have been delighted to see its main feature preserved for posterity, whilst the chance to build the railway was a proposition difficult to resist, despite his considerable ongoing commitments. There were also hard-headed business reasons for supporting the WEL&CPR. Besides likely profits from the works, the new approach to the capital from the south was likely to make the railway a desirable acquisition for the existing companies, in particular the Brighton Railway, which had to be content at this time with sharing the facilities at London Bridge with the South Eastern.

Although Brassey again joined Peto and Betts to undertake the WEL&CPR, there never appears to have been any question of the contractors taking a lease of the new line, as they had the LT&SR two years before. They either no longer wished to accept the liabilities, or more likely considered the company capable in the improved economic climate, of raising the capital it needed without such an incentive. Betts did however sign the subscription contract drawn up before the scheme was submitted to Parliament, agreeing to invest £60,000, which made him the largest investor at that stage.[26]

The price agreed for the construction of the line was £310,000, excluding the cost of land and the stations, which worked out at nearly £40,000 per mile. Although two tunnels were needed, this exceedingly high figure was hardly justified by the work involved. As a letter from Bidder to the WEL&CPR directors early in 1854 suggests, the cost was probably a direct consequence of the boom conditions prevailing at the time:

> In reference to the offer of Messrs. Peto, Brassey and Betts … considering the uncertainty of the labour market and the price of materials and seeing that already since the parliamentary estimate was framed, a rise of fifteen per cent has taken place in these respects, I am of the opinion that it will be provident to accept the … tender.[27]

Although these were words which Peto might very well have used, the extent to which Bidder's judgement was influenced by him cannot be known. Besides wishing to maximise the profits on the works, Peto also had to take into account any further support he might be obliged to give the company before the contract was completed.

The WEL&CPR soon experienced difficulty in raising the large sum needed and the outbreak of the Crimean War in March 1854, made matters worse, by depressing the investment market. A month after the commencement of hostilities, £100,000 was still needed to cover the cost of the first section of the line under construction, from Sydenham to Wandsworth Common. This was not completed until the end of 1856, the delay being due in part to the need to realign the tunnel which passed underneath the Palace grounds, to avoid undermining the foundations of the water tower. By this time the contractors had found themselves obliged to accept £80,000 of debentures in lieu of cash, although following the precedent set with the LT&SR, they insisted this was made over to them 'in anticipation of the works'. They were soon called upon to provide further assistance, with Peto subscribing £112,000 WEL&CPR 6% preferential stock and arranging for loans of £325,000 from the Union Bank and £80,000 from the Brighton Railway. He also agreed to dispose of all the forfeited shares. The directors recognised their indebtedness by giving him a seat on the Board on the completion of the contract in 1858; proceeding to elect him to the Finance Committee, where he was instrumental in obtaining a further loan from the Union Bank and another from the Rock Life Assurance Company.

The Brighton Railway had given the fledgling WEL&CPR its support, because it wanted to use the new line to obtain access to the West End. Initially an end-on junction was made at Sydenham with the short spur line the Brighton had built earlier. This was soon supplemented with a link to Norwood, which allowed its trains to run directly onto the WEL&CPR. With the Brighton also agreeing to operate the WEL&CPR, the first steps had been taken towards the amalgamation, which had very likely been in Peto's mind from the time he first became interested in the scheme. He proceeded to use his position on the Board to spearhead the negotiations with the Brighton, which culminated in its acquisition of the company in 1860. The terms were good for the WEL&CPR, holders of its 'A' shares being guaranteed 4% interest, those with 'B' shares 7%, whilst the preferential stock, much of which had by then come into the possession of the contractors, continued to carry interest at 6%.

Meanwhile a new company had been promoted to extend the WEL&CPR line over the Thames to a new terminus at Victoria, but the WEL&CPR's involvement with this, prevented Peto from undertaking the work. In 1856, probably in order to leave the way clear to join the WEL&CPR Board, he had also deprived himself of the opportunity offered him to build an extension from Sydenham to join the Mid Kent Railway at Farnborough. He was instrumental in this contract being placed with Messrs Smith & Knight, whose prices were in his view, not only fair but 'below what he himself would have accepted'. He may well have had his own, as well as the company's interests in mind, when he gave this advice to the Board , as two years later he would take over the contract for the Severn Valley Railway (SVR) from Smith & Knight. It is not known what ties, if any, Peto had with this firm.

With a view to obtaining another contract, Peto had also backed the Brighton company's scheme for a branch to Caterham. This had the additional attraction of being linked with a proposed housing development in the embyronic suburb, designed to appeal to better-off City clerks wishing to move out of London. Although aimed at a less well-off clientele than the slightly later venture at Southend, it was also designed to provide revenue for the railway, even if the journey to work was shorter. The Brighton company had originally backed the scheme but once the railway came within the orbit of the South Eastern, Peto's links with its rival meant he had little chance of obtaining the contract.

THE WIMBLEDON & CROYDON RAILWAY

PETO was only to build one line besides the WEL&CPR on the southern periphery of London in the 1850s. This was the Wimbledon & Croydon Railway, which was promoted by an independent company and utilised a portion of the by then defunct Surrey Iron Railroad. As well as being the engineer, Bidder agreed to lease the line on completion, and may also been responsible for the contract being placed with Peto & Betts.

The six mile line was built for £45,000, making it considerably cheaper than the WEL&CPR, but it was shoddily constructed, the Board of Trade Inspector twice refusing to sanction the opening. Worse followed a few weeks after services commenced in 1855, when defective ballasting caused a serious derailment at Mitcham that killed the driver of the train involved. Although one of his smallest contracts, it was a significant blot on Peto's reputation for reliable workmanship. Bidder for his part was probably relieved to be rid of the lease after only two years.

THE EAST SUFFOLK RAILWAY

MEANWHILE, back in East Anglia, having resigned the chair of the Norfolk Railway in 1855, Peto was free to proceed with his cherished ambition of providing Lowestoft with a new more direct rail link with London than the existing route through Ipswich and Norwich, provided by the Eastern Counties and Eastern Union companies. The nucleus of the project was the small independent Haddiscoe, Beccles & Halesworth Railway, which had opened a branch running southwards from the Lowestoft line in 1854. Peto was the principal shareholder in this and proceeded to propose that it be extended along the coast to Woodbridge, to join a branch, which the Eastern Union company was proposing to build from Ipswich. The new company, to be known as the East Suffolk Railway (ESR), was incorporated in 1856, to include extensions from Beccles to Lowestoft and from Haddiscoe to Yarmouth, as well as branches to Leiston, Snape and Framlingham. Peto offered to pay the shareholders 3½% interest on their capital whilst the line was being built, and then to lease it for twenty-one years at 6%.

The Eastern Counties Railway (ECR) had a vested interest in the Norwich route and opposed the project from the start. This presented a problem for Peto, because the ECR's London–Colchester line formed an integral part of his new route to the metropolis. His response was to promote the Pitsea, Maldon & Colchester Railway to link the LT&SR – which of course he already leased and effectively controlled – with the EUR, and free him of any dependence upon the ECR. The new scheme may have been a bluff but the ECR took this threat to its position very seriously. Its directors were probably behind the publication of a hard hitting attack on Peto and his business methods in the satirical publication entitled *Petovia: being a review of the scheme for a railway from Pitsea to Colchester, and an exposure of the motives which promoted it, the absurdities that characterize it, and the inevitable failure that awaits it*, which appeared in 1857. The author of this, who used the pseudonym 'A Tooth of the Dragon', claimed that by charging £15,500 per mile for the line, when it could be constructed for £8,000, Peto and his partners would make sufficient profit from the works to cover the cost of the lease. He added that the other objective of the scheme was to divert traffic from the ECR to the LT&SR, to reduce the contractors losses on that lease. *Petovia* also exposed some of Peto's other failings as a contractor, including the rotten state of much of the woodwork he had used on the Norfolk Railway.

The directors of the ECR continued to criticise Peto when the ESR proceeded with its Bill. Echoing the comments expressed earlier about the Pitsea project, they castigated him over the high cost of constructing the railway, which had risen to £15,000 per mile, after he had agreed to lease it. Earlier, without this guarantee, a figure as low as £9,000 per mile had been suggested, although this had not included laying double track now proposed on the whole of the main line south of Beccles. The ECR claimed the railway could be built for as little as £7,000 per mile, saying that at £11,000 per mile the contractors stood to make an overall profit of £248,000 – a sum which, invested over the period of the lease, would net them nearly £500,000.

In the end a compromise was reached, Peto abandoning the Pitsea scheme, in return for the ECR withdrawing its opposition to the ESR.

So the bluff, if such it had been, had paid off. But in deference to the criticisms which had been directed at him, Peto left the construction of the link line from Woodbridge to Ipswich to his less controversial partner, Brassey who entrusted it to a separate engineer, Mr Boys; whilst Bidder had charge of the construction of the remainder of the ESR. This did not prevent this ten mile section costing £144,000, only fractionally less per mile than the rest of the railway, which Peto proceeded to build jointly with Betts and Brassey.

Bidder economised on the construction of the ESR by choosing a route which avoided crossing the river estuaries along the Suffolk coast. The result was a line which, by East Anglian standards, was rather steeply graded and contained a number of sharp curves, whilst the flat country-side through which it passed was responsible for the proliferation of level crossings, which would add appreciably to working costs over the years. The cheapness of the works benefited the lessees in the short term and also enabled the line to be built quite quickly, reducing the period during which they had to pay interest to the share-holders without receiving any traffic receipts.

The railway was opened in its entirety on 1 June 1859, Brassey having co-ordinated the work on the somewhat more difficult Ipswich to Woodbridge line, to have this ready on the same day. There were no formal celebrations until 18 July the following year, when the directors got round to organising a banquet at the Royal Hotel Lowestoft, which had recently been opened as part of Peto's extensive developments on the south side of the town (*plate 17c*). To obtain the maximum publicity for the ESR and the town, he arranged for a full account of the proceedings to be published. An extract from this gives the flavour of the day:

> From an early hour on the day of the banquet, the town of Lowestoft, especially in the immediate vicinity of the Royal Hotel, presented a cheerful and pleasing appearance, the flag-staff being gaily decorated with a variety of flags, while a more than usual number of ladies and gentlemen crowded the Esplanade, the Pier and their approaches.[28]

The banquet in the evening was presided over by the Earl of Stradbrook, the Lord Lieutenant of Suffolk, who had supported the ESR scheme, and other local dignatories, as well as the directors of the railway. The event turned out in the end however, to be as much a eulogy of Peto as a celebration of the completion of the ESR. At the head of the table stood a dessert service and some plate, the centre piece of which was 'supported on a separate plateau, richly chased … with large shields on either side' presented to Peto by the Earl and 'upwards of three hundred other subscribers connected with East Suffolk, as a token of personal regard and as a grateful expression of the obligation they are under to him for making the East Suffolk Railway'. After the usual speeches and junketing, the assembly was treated to a rendition of a song written for the occasion by William Day, one of the directors of the company. Notwithstanding the doggerel verse, this is worth repeating in full, as evidence of the fulsome praise lavished upon Peto at the height of his power by some railway investors.

> *A hearty welcome let us give*
> *Our worthy friend and guest*
> *And long amongst us may he live*
> *With every comfort blest*
> *To public usefulness he hath*
> *Devoted all his days;*
> *And he who toils in such a path*
> *Deserves the public praise.*
>
> *Our railway gives to him a claim*
> *To this our full assent;*
> *For that has raised up to his fame*
> *A lasting monument:*
> *A work for which we owe him much,*
> *Much more than we can pay;*
> *But we may show our thanks are such*
> *As grateful men display*
>
> *Then charge all glasses to the brim*
> *With bright and cheering wine*
> *And drain them to the health of him*
> *To whom our hearts incline!*
> *With three times three, and one cheer more*
> *The toast aloud proclaim,*
> *And let this festive hall resound*
> *Sir Morton Peto's name.*[29]

Peto treated his admirers to a lengthy oration, in which he had no inhibitions in acknowledging his own 'very large responsibility in connection with the undertaking' but he felt it necessary also to give a few words of reassurance to those who had invested in the enterprise. Only too aware of the problems the LT&SR was by now experiencing, he told them that rural lines like the ESR needed to operate for at least three years before they could realise their full traffic potential. Although the ESR shareholders, unlike their LT&SR counterparts, had no right to share in any profits, Peto having been careful to avoid writing any such provision into the terms of the lease, they were nevertheless concerned about the market value of their investments.

The occasion marked the high water mark of Peto's activities in East Anglia. He would build no more railways in the region and within three years would dispose of the Somerleyton estate, which had become something of a burden. Although few outside his immediate circle would have realised it at the time, he was also in more straitened financial circumstances. This was less the result of his heavy commitments to railway projects at home, as the disastrous losses he and Brassey had suffered from a single foreign contract, the Grand Trunk Railway of Canada. They had been involved in that fateful enterprise before the ESR was even conceived, and were fully aware how much it had cost them by 1855. It is some indication of Peto's implacable self confidence that once he had weathered the Canadian storm he went ahead straightaway with the ESR project, when a more cautious entrepreneur would have held back. Putting aside the question of his prestige in the locality at the time, he might well, gambler as he was, have seen the new venture as the best means of restoring some of his losses.

The gamble seems, in part at least, to have paid off. Despite the ESR's doubtful prospects it did find a purchaser within a comparatively short time, being absorbed in 1862, on favourable terms for its shareholders, into the recently created Great Eastern system. The ESR shareholders received GER stock in exchange for their holdings in the former company and at the same time the lease to the former contractors was terminated. So, whilst it cannot be known how Peto fared personally at the end of the day, unlike the LT&SR, the ESR did not become a long-term liability. He was also relieved of £86,000 the ESR owed him for the works.

THE SEVERN VALLEY RAILWAY

WHILST in the midst of executing the works in Suffolk, Peto agreed terms with the Severn Valley Railway (SVR) to build an extension from the Oxford, Worcester & Wolverhampton Railway (OW&WR), which he had completed four years earlier. As far back as 1851, when work had only just begun on the OW&WR, he had been approached by the original promoters of the scheme – the provisional Chairman, Major Tyndale, hoping for substantial financial assistance as part of 'an arrangement for the execution of the works'. Peto decided not to enter into any formal agreement at that time however, although he, and subsequently Brassey, agreed to invest in the scheme. This was sufficient encouragement for the Provisional Committee to go to Parliament and obtain an Act for the line in 1852, but investors were slow in coming forward, even in the locality the railway would serve. By this time both Peto and Brassey had enough contracts in hand not to wish to take on another, especially one involving substantial commitments on their parts. Probably aware of the level of support their prospective contractors were giving at the time to the neighbouring HR&GR, the SVR promoters tried to pressurise them to make an early start on the SVR, but soon found out who was in the stronger position. At the end of 1853 Peto told them:

I am anxious that no expense of any kind should be incurred until we are really in a position to proceed. You have abundance of time and if we are kept from beginning for the next three years, we can then open it long before the Act requires. I need not impress upon you the present state of the money market or the political condition of Europe.[30]

When the promoters declined to follow his advice, seeking parliamentary approval for a revised scheme with reduced construction costs, there was a heated exchange of letters; followed by a threat from Peto, not only to withdraw his support, but also to oppose the proposed Bill when it came before Parliament. The Committee had no option but to withdraw the Bill but even this did not satisfy Peto. Determined to exert his authority and nip any independence on the Committee's part in the bud, he insisted he be given a seat on the Board. This shows how far he had come in the twenty years since being browbeaten by Hammond at Hanwell, and how he could be as ruthless as any company chairman when the situation demanded. He got himself duly elected Chairman of the SVR.

During the three years Peto was Chairman of the SVR, he virtually dictated the financial policy of the company. In return for additional financial support and agreeing to act as trustee for all the forfeited shares, the Board authorised him to raise funds 'as he thought fit'. He responded by insisting that income from 'calls' upon the shares should, in the first instance, be used to pay off the advance he had made to enable the SVR to cover the deposit required to obtain its Act.

Another Act was obtained in 1855, the main purpose of which was to further reduce the capital cost of the project. Peto's support for this suggests his antipathy towards the earlier proposal was not because it would make the contract less profitable, but had everything to do with its timing *vis-à-vis* his other commitments. The new Act reduced the cost of the railway from £600,000 to £480,000, mainly through some economies in its construction, including the shared access to Shrewsbury with the Hereford & Shrewsbury Railway.

After another Act had been obtained in 1856 covering share issues, Peto felt able to proceed, but as Chairman was of course prevented from undertaking the work himself. In July 1857 the Board agreed to place the works in the hands of Messrs Smith & Knight, provided this firm's tender was acceptable to the SVR engineer, John Fowler. Fowler had replaced the original engineer, Robert Nicholson, who had died in 1855. Within a fortnight of the offer being made

to Smith & Knight, however, Peto informed the Board he would resign as Chairman and take on the contract himself. This was the reverse of what happened with the Farnborough Extension contract, referred to earlier (*see p.52*).

As part of the contract deal, Peto and Brassey agreed to subscribe £225,000, nearly half of the total share capital of the company, and to accept shares in lieu of cash for the works up to a maximum of £240,000, which meant that, as earlier with the Hereford company, they would be able to meet the 'calls' through the work they carried out. The commitment was nevertheless formidable, the initial subscription being almost double what they had provided for the HR&GR. In return the SVR agreed to pay the contractors £363,000 to construct the line, excluding land, stations and legal charges. This worked out at about £13,000 per mile, £3,000 per mile more than the Hereford line but slightly less than the East Suffolk. The price was not excessive, considering the nature of the terrain the line had to cross.

Difficulties in raising the rest of the capital on the open market delayed the start of work and made Brassey at one stage despair of the whole venture, informing the SVR directors that he had 'no great confidence in the ardour of parties, who can only be moved by meetings for cutting the first sod'. By this time he had a larger stake in the company than Peto, probably because his organisation would be carrying out the work. The SVR could perhaps have been a *quid pro quo* for Peto's execution of the Hereford Ross & Gloucester works. Brassey's agent was certainly in charge of operations and he was consulted more often by the SVR directors during the construction period than had been the case with his other joint undertakings with Peto.

Although the SVR followed the course of the Severn for nearly all its route, from the junction with the OW&WR at Hartlebury, just north of Worcester, to Shrewsbury, it was not an easy line to construct. As well as tunnels at Bewdley and Bridgnorth, a major viaduct was needed to cross the Severn at Arley. Fowler designed this as a single cast-iron arch, with a span of 200 feet, making it the largest work of its kind in Britain up to that time. The heavy nature of the works, problems in purchasing the land together with an exceptionally wet summer in 1860, all contributed to the delay in completing the line. Brassey had it ready for traffic in January 1862, nearly four years after the contract had been signed.

The SVR proved another case where traffic did not match the expectations of the promoters. Although the Great Western Railway agreed to work the line from the start, it would be ten years before an amalgamation could be effected with the West Midland company. In the intervening years the SVR was obliged to struggle along as an independent concern, with its shares in a depressed state. It is not known how quickly or on what terms Peto and Brassey disposed of their substantial holdings in the company.

A CONTROVERSIAL FINANCIAL STRATEGY

DURING the 'fifties, with significant assistance from Brassey, Peto had made himself a major railway promoter, whilst still retaining the outward vestige of a reliable, well organised and resourceful large-scale contractor. His original role was now the perfunctory *raison d'être* for an increasing complex business enterprise, with the eventual profit from the works becoming more subordinated to the investment potential. The financial manipulations linked with the promotions had made him a controversial figure in the railway world by the end of the decade, his methods being succinctly and vituperatively described in *Petovia* as:

> The concoction of companies by contractors, in order to make their own terms for construction of the works; or the bribery of directors … to the sacrifice of the shareholders; and, in either case, the contriving to get overpaid for work done, at the expense of the unfortunate, deceived, and plundered proprietary.[31]

Elaborating upon the same theme, the author, whoever he was, asserts that 'Petoism' …

> has long surrounded our joint stock companies with its pestiferous miasm,

and can no longer be tolerated … Morally dishonest, socially destructive, it may or may not be legal thus to sacrifice the interests of thousands at the shrine of self and pelf. But the evil can be no longer endured; and the shareholders … who shall from this time forth regard with indifference, the attempts of conspirators against their best interests to wriggle themselves into their confidence and into power, that they may sell them or plunder them, will deserve the consequences they court.[32]

The sums of money needed to sustain the various promotions must again call into question Peto's ability to maintain this level of financial commitment, even allowing for the substantial assistance presumably provided by Brassey. His own strongest asset throughout the period was undoubtedly his creditworthiness, not only in the eyes of the City institutions who were backing him, but the significant number of ordinary shareholders, who invested in the lines he leased on the strength of his personal guarantees. Whilst enough people appear to have thought that Peto's credentials were impeccable, there were some who either had genuine doubts, or, as in this case of the author of *Petovia*, wished to discredit him, by spreading the rumour that he had seriously over-extended himself, saying:

> I consider him the railway gambler par excellence – the successor of the notorious Hudson – in the wholesale extent of his undertakings, the ingenuity with which he cooks prospectuses, in the magnitude of his audacity and the success with which he angles for dupes. But the pitcher which often goes to the well is tolerably sure to be broken in the long run and the real friends of the Baronet are they who caution against the madness of playing double or quits.[33]

This was sound advice, which he would have done well to heed.

SIX

The London, Chatham & Dover Affair, 1860–1868

ORIGINS OF THE SCHEME

AFTER nearly thirty years in business and a quarter of a century as a railway contractor, Peto might have considered retiring when the 'sixties dawned, as his former partner, Thomas Grissell, had done several years earlier. Barely fifty, in robust health, and still displaying the energy and self-confidence which had taken him so far in the world, he had no intention to bow out for the time being. Underneath the façade all was not well, however; his ongoing commitments coming on top of the losses abroad, may well have required another financial coup on a par with the Mania bonanza, to secure final prosperity.

The successful outcome of the West End of London & Crystal Palace Railway (WEL&CPR) venture had probably confirmed Peto in his belief that his future as contractor and promoter, lay within the metropolitan orbit, although the Caterham setback and the enforced abandonment of the Farnborough Extension, diverted him back to provincial works for a while. Before the 'fifties had passed, however, he obtained a contract from a company which was planning to challenge the railway *status quo* south of the Thames. The East Kent Railway (EKR), originally set up to build a branch off the South Eastern line at Chatham to Faversham, proceeded with extensions eastwards to Dover, and in the other direction to join the Mid Kent Railway (MKR), which in conjunction with the Farnborough Extension and the WEL&CPR, would give it access to the new West End terminus at Victoria. In 1858 the EKR engaged Peto to construct the whole of its new line from Strood, just across the Medway

from Chatham, to a junction with the MKR at Bickley. The price was high, the twenty-four miles of double track costing £738,000, or just over £30,000 per mile, suggesting that there may have been financial commitments attached, although Peto later denied this. Peto embarked upon this undertaking again with Brassey but he soon withdrew, deciding, as it turned out very wisely, to sever all connection with the EKR. Peto promptly found a new partner to join him and Betts. Thomas Crampton was a leading locomotive designer, who had expanded his activities into contracting, being engaged at the time in constructing the EKR's Dover extension *(plate 18b)*. The Peto, Betts and Crampton partnership completed the Bickley line in 1860.

Before the Bickley line was finished, the EKR, having in the meantime renamed itself the London, Chatham & Dover Railway (LC&DR), had embarked upon another, even more ambitious project, designed to give it independent and more direct access to London, than was provided by the somewhat circuitous and steeply-graded WEL&CPR route. Under the Metropolitan Extensions Act obtained in October 1860, the LC&DR was authorised to build an extension from Beckenham, on the Mid Kent line, to Herne Hill, and from there to launch a two-pronged attack upon the capital, with a 'Western line' running roughly parallel to the WEL&CPR to end, like it, at the new Victoria terminus, and a 'City Extension' proceeding due north to cross the Thames at Blackfriars and join the Metropolitan Railway (MR) at Farringdon Street.

The scheme involved expensive engineering works. A tunnel 2,141 yards long was needed at

Penge on the Beckenham–Herne Hill section, whilst the 'City Extension' required four miles of brick viaducts on its approach to the Thames, a bridge over the river at Blackfriars *(plate 18c)* and a tunnel under Snow Hill in the City. To improve access to Victoria from the 'Western line', it was also proposed to widen the existing river crossing at Pimlico. The costly construction was compounded by high land values, particularly in the City and the need to demolish a considerable amount of property, especially between Herne Hill and Farringdon Street. The LC&DR, in consequence, needed a great amount of capital to realise its ambitions, being authorised under the provisions of the 1860 Act, to raise £1,750,000 in shares and to borrow a further £583,000. This was for just twelve miles of track, which worked out at over £200,000 per mile, and made it the most expensive overground railway to date in Britain.

Peto did not initially enter into any formal arrangement to assist the LC&DR, although he may have offered some informal guarantees, as the company placed the contract with him without seeking other tenders and also agreed to pay for the works on a schedule of prices. The appointment of Joseph Cubitt as engineer could also be seen as a sign of Peto's influence, as, besides working together on the Bickley line, they had also been involved with the Great Northern scheme. The LC&DR for its part, was happy to hand Peto £15,000 of debentures and £45,000 of preferential shares, just after he began the construction of the Metropolitan Extensions.

Although the scheme was launched during an investment boom, Peto must have had doubts from the start as to how much of the massive capital sum could be raised directly through share issues. Evidence for this is to be found in the attempt he unsuccessfully made earlier to persuade the LC&DR to opt for a cheaper route through Sydenham. His fears were soon proved justified. Within a year of work starting, the LC&DR found that the capital provided for in the Act was insufficient, and matters were made worse because in the meantime the company had embarked upon an additional project in collaboration with the Brighton Railway. The two companies agreed to build separate portions and then share use of the South London Railway (SLR), which was planned to link Victoria with the Brighton's other London terminus at London Bridge, by means of a new line through Brixton and Peckham Rye. This would also connect with the LC&DR's 'City Extension' at Loughborough Junction and allow the company to build an 'Eastern Extension' from Peckham Rye to Greenwich. The approach to Victoria was also to be further improved by means of a new high level line from Factory Junction in Brixton to Grosvenor Bridge and the widening of the bridge over the Thames.

In 1862 the LC&DR was authorised to raise a further £1,050,000 for the additional lines but when this proved insufficient had to obtain another Act two years later, which allowed it to raise a further £1,600,000, nearly half of this sum being in loans. The 1864 Act also permitted the company to make the most expensive part of the 'City Extension' north of Blackfriars Bridge a separate undertaking, to be known henceforth as the 'City Lines' and to raise £1million specifically for this. The Victoria approach and the 'Eastern Extension' were also separated from the Metropolitan Extensions proper, with provision for £370,000 and £850,000 respectively to be raised to meet their cost. In all the LC&DR was now commiting itself to an outlay of nearly £8¾million, a gigantic sum for a comparatively small and fundamentally weak company. The contracts for all the new works, including the LC&DR's portion of the SLR, were placed with Peto and his partners, again without any other tenders being sought.

PETO TAKES CHARGE OF LC&DR FINANCES

THE escalating cost of the project and the company's inability to raise enough of the capital on the open market inevitably led to requests for assistance from Peto. Although he accepted bonds and debentures in lieu of cash for the works from an early stage, it was not until 1863 that he allowed his name to be directly linked with any of the LC&DR share issues, when he agreed to guarantee the interest payments on Metropolitan Extension B shares. In return the company, like the WEL&CPR in similar circumstances a decade earlier, formally expressed its gratitude by making him its Financial Adviser, with the right to attend all Board meetings. In the case of the LC&DR, however, Peto was assuming the position at a much earlier stage, when its works were by no means completed, which inevitably involved him in raising the large sums of money needed to complete the project.

One of Peto's first steps after becoming Financial Adviser, was to persuade the LC&DR directors to agree to a change in the terms of his contract for the original 'Metropolitan Extensions'. All the payments due to him under the original terms were consolidated into a single fixed-price contract of just under £6million, which included the cost of land, legal charges and interest due to the existing shareholders. By agreeing at the same time to accept payment for all but £½million of this sum in LC&DR securities, Peto was in effect underwriting the largest part of the whole project. Never before, in Britain at least, had he been personally, or even jointly, so deeply committed to any single company. The LC&DR for its part, although relieved of the need to raise the capital involved, would pay dearly for the privilege. The total cost of the scheme had more than trebled since its inception, with the inevitable result, as with all overcapitalised enterprises, of excessive debt on the capital account and all the consequent long-term liabilities.

The LC&DR had already been lumbered with a considerable volume of debt when it embarked upon the Metropolitan Extensions, most of this being attributable to its earlier schemes. Some £1¼million of liabilities took the form of 'Lloyd's bonds', a comparatively new type of security, which had been introduced onto the market by the barrister John Horatio Lloyd. They were designed to meet the needs of newer and smaller railway companies like the LC&DR, which, in order to raise large sums of money quickly, were obliged to offer incentives to attract investors. Lloyd's bonds carried higher interest than ordinary shares and also had first call upon the company's resources, before even debentures,

as far as their repayment was concerned. The size of the LC&DR's liabilities to its Lloyd's bondholders made it more difficult than it would have been otherwise, for the company to raise new capital through ordinary share issues, and also depressed the value of its debentures. As the major future recipient of shares, Peto was naturally concerned on both accounts but especially did not want the debentures to be allowed to depreciate too far, as these were the most readily realisable form of scrip, as far as he was concerned. There was therefore a considerable element of self interest involved when Peto told the LC&DR directors in 1863, that they could not 'go into the market as honest men to borrow money' without first paying off all the outstanding Lloyd's bonds.

The means Peto proposed to pay off the bondholders, in due course formally agreed to by the LC&DR directors, was to create a special 'Consolidated Common Fund' consisting of £1,500,000 of new stock and £800,000 of unsold old Third Preference shares, all of which would be charged against the company's General Undertaking as distinct from the Metropolitan Extensions and related accounts, the revenue from this new source to be used to redeem all the outstanding bonds. The entire 'Consolidated Common Fund' was apportioned amongst Peto, his partners and staff, and one of the LC&DR directors, Sir Cusak Rooney. None of these persons was required to pay anything upon what was, for the time being at least, an entirely nominal holding, nor was any attempt made to place any of the stock on the market. Instead Peto used the entire issue as security for the loans he proceeded to raise from a number of banks and discount houses, giving these institutions his personal guarantee that their cash advances would be duly re-paid. When this ploy was later exposed for what it really was, a blatant instance of pyramid financing, Peto could only justify his actions by claiming it had been rendered necessary because the LC&DR directors had refused to sell him the Consolidated fund at a substantial discount.

The effect of Peto's manipulation of the company's accounts had been to nearly double the original bond capital by re-mortgaging the old liabilities and at the same time unsold shares had been refloated, disguised with new guarantees from the contractors. Whilst the LC&DR stood to lose most in the long term through the liability for interest on its inflated share capital, Peto and his partners in the shorter term were obliged to meet the charges on the loans as these became due. Although guaranteeing the Consolidated stock was in theory a much greater liability, there seemed little likelihood of having to redeem it at the time. The risk was certainly no greater than that taken by those who provided the cash loans on the security of, what was really unsubstantiated collateral.

Despite opposing the discounting of the Consolidated Fund on Peto's terms, the LC&DR directors were prepared from quite early in the construction to pass over large quantities of Metropolitan Extension shares to him at substantial discounts. He stood to gain personally from their subsequent sale, although he could never be certain he would be able to dispose, especially of the ordinary shares, at a higher price than he had paid for them. The discounting of shares was to prove a matter for dispute between Peto and the company later, but at the time those purchasing the shares were unaware of the cut he was sometimes taking. At the end of the day on the company's reckoning something in the region of £2½million of the share issues was effectively written off in this way. This is not to say, of course, that the shares in question would not have had to be discounted had Peto not handled them. The suspicion remains however, that he did rather well out of at least some of these transactions.

Well before he had completed the Metropolitan Extensions, as now redefined, Peto was finding it increasingly difficult, despite discounting the shares, to raise the large sums of ready cash he needed to purchase land and construct the line. Consequently he was obliged to resort to more devious means of raising funds. One was to mortgage some of the assets accruing from the construction of the railway, especially surplus land, which was fairly easily realisable. In at least one instance land set aside for future use by the company, was mortgaged. *The Economist* was subsequently able to cite an instance of this happening in 1864, the land in question having been set aside by the LC&DR for a future goods depot at Blackfriars:

> Upon the security of these alleged lands, Sir Morton Peto arranged … for the advance of £135,000 from the Imperial Mercantile Credit Association and a further sum of £190,000 from the General Credit and Finance Company. These two companies were not to advance any actual money at all. They were to give their acceptances to bills to be drawn from time to time by 'firms or individuals resident on the Continent of Europe' … the bills so drawn were to be handed to the borrowers for discount. The credit companies were to be paid a commission of five per cent for the twelve month … and in the untoward event of the finance companies being at any time under actual cash advance, the rate was to be four per cent above the minimum Bank rate, such never to be computed … at less than eight per cent; so that for the use of actual money, the borrowers were willing to pay twelve per cent and five per cent commission.[34]

These were exorbitant rates to pay at the time but, as *The Economist* proceeded to point out, Peto did not expect to be called upon to pay more than the commission on the bills, which he relied upon being able to cash in the City on his own creditworthiness. He could rely upon a number of City discount houses, of which Overend and Gurney was certainly one, to cash such bills, if and when he needed money to meet the ongoing cost of building the railway. The Imperial Mercantile Credit and the General Credit and Finance, which fulfilled the essential intermediary function on this occasion, were fringe banks, of which a number had sprung up in recent years, who attracted 'hot' money by offering above average rates of interest and advanced loans on the more speculative type of enterprise. Peto was a major shareholder in the Imperial Mercantile, which no doubt smoothed the way for this transaction. *The Economist* had no hesitation in condemning the whole arrangement as bogus, going on to assert that the individuals resident on the Continent of Europe had really been 'boys in offices signing bills at a few shillings a quire and quite unable to back a loan of £325,000'. As if this pyramid of inadequately secured credit was not enough in itself, it noted that a loan was to be obtained from the Rock Insurance Company, on the security of the site when the goods depot had been completed.

FINANCIAL CRISIS

THE financial situation had not prevented Peto from making steady progress with the construction of the Metropolitan Extensions, the 'Western line' being completed from Beckenham to Stewart's Lane in Brixton by July 1863 and the 'City Extension' carried as far as a temporary terminus on the south side of Blackfriars Bridge by the end of the following year. Efforts were then concentrated on the bridge itself, which had to be finished by December 1864, to comply with the conditions set by the City authorities. In the final hectic stages of its construction more than 3,000 men and boys were working there. Designed by Joseph Cubitt, using the American truss system, the work had consumed more than 3,000 tons of iron and 600,000 rivets by the time it was finished.

There remained the 'City Lines', running from Earl Street, just north of the bridge at Blackfriars, to Farringdon Street, which was the most expensive stretch of track in proportion to its length of the whole scheme, involving an outlay of £1½million for less than two miles of track (plate18a). Most of this was accounted for by the high cost of land and the need to demolish a considerable amount of property, the works themselves only costing £193,000. With the exception of £300,000 subscribed by the Great Northern Railway (GNR), which in conjunction with the Metropolitan Railway (MR) would make use of the new north–south route across London, the LC&DR had left Peto to raise all the capital needed. This was a formidable task at the best of times but particularly difficult when money became increasingly tight, as the early

'sixties boom subsided and the economy came under increasing inflationary pressure. Peto might well have considered suspending operations had the 'City Lines' not been so crucial for the profitability of the whole venture, by providing all-important access to the City, as well as considerable potential revenue from the other railway companies wishing to use it. There was little room for economies in the construction, as some of the more expensive embellishments, including an ornamental arch at the foot of Fleet Street, had been incorporated to appease a powerful lobby in the City that had been hostile to the railway from its conception. It was, however, decided at a fairly late stage to abandon a proposed branch to Smithfield Market.

INSOLVENCY

PETO pressed ahead with the 'City Lines' in increasingly difficult circumstances through 1865 into the early months of 1866. By this time the country was in the grip of a major economic crisis, brought about mainly by the excessive speculation of the previous five years, some of which was attributable to ambitious new railway schemes like the LC&DR. The collapse of the markets was initiated, however, by the political tension in Europe that preceded the outbreak of the Austro-Prussian war in the summer of 1866. As investors withdrew their deposits and money became tighter, the first to feel the effects were inevitably the more speculative companies and the institutions that had backed them. By April the list of bankruptcies was lengthening alarmingly, with at least one major railway contractor, Thomas Savin, obliged to file his petition. Very few in the City were prepared, however, for the events which were to follow in the next few weeks.

The crisis came to a head on 12 May, long remembered in the City as 'Black Friday'. The panic in the financial markets had been triggered two days earlier with the collapse of Overend and Gurney, the discount house for long regarded as a pillar of the City. Overends had largely brought their troubles upon themselves, having been badly managed over a number of years, advancing cash upon doubtful securities and being excessively committed to speculative ventures. Peto, as mentioned earlier, had been a customer for a number of years, although it is not possible to know how extensively he made use of its services, or whether his activities played a significant part in bringing the firm down. With Overends demise he was deprived of his last large source of ready cash and it is an indication of the hand-to-mouth state of Peto's finances by this time, that only two days after Overends closed their doors for the last time, his own firm's name was added to the long list of those obliged to suspend payments.

For Peto the irony was that he had by this time completed virtually all the redesignated Metropolitan Extensions, as well as the costly 'City Lines'. Indeed all that remained unfinished of the whole project, was some track widening near Farringdon Street and on the approach to Victoria, part of the LC&DR's portion of the SLR and the 'Eastern Extension'. It seemed possible that he might be able to complete these works if his difficulties were only temporary, as he insisted they were when he addressed a meeting of his creditors on 8 June. According to the balance sheet he presented that day, his firm had assets of more than £1½million to set against liabilities of only £430,000, although closer scrutiny of the document would have revealed that he only had about £8,000 cash in the bank; whilst his only other immediately realisable asset was his plant, which was estimated to be worth no more than £50,000. Nevertheless he was given a sympathetic hearing at the meeting, most present apparently believing him, when he proclaimed that he could not only meet all his liabilities but would, in addition, pay interest on what in the meantime he owed his creditors. He also insisted that he would have survived the recent crisis, if he had accepted the loans that had been offered him at the time, but had not wished to risk 'bringing down tried and trusted friends'. How true this was can never be known, or indeed, who the 'trusted friends' were. One of these could well have been Brassey, who, although not involved in the LC&DR's affairs, still had ongoing commitments with

Peto, such as the London, Tilbury & Southend Railway, as well as other undertakings abroad, and therefore had an interest in his survival. Brassey had been under considerable pressure himself during the crisis, however, and there would have been limits to any assistance he might have provided to Peto.

Not surprisingly, the LC&DR suffered severely as a result of Peto's insolvency, its finances being so closely intertwined with his. Deprived of funds from the accustomed source and with recently-completed lines only just beginning to earn revenue, the directors found themselves unable to meet the interest payments due upon the company's debentures in 1866. The shareholders' immediate reaction was to set up a Committee of Investigation, which was charged with the task of looking into all aspects of the financing of the recent schemes. This boded ill for Peto, but he had some time in which to gather his thoughts and prepare his defence, as the shareholders' venom at this stage was directed mainly at the Board, who were castigated for permitting the project to be so overcapitalised and wasting the company's resources, by discounting shares and paying excessively-high interest charges.

The Committee of Investigation published its report in October 1866. This revealed irregularities in the financing of the scheme that involved Peto so intimately, that he now knew he would be at the centre of the row that was about to break out, with no possibility of hiding behind a mantle of contractual innocence. His only hope was that the LC&DR directors had been party to most of his actions, and consequently were obliged to stand by him. This proved to be the case – their initial response to the Committee's findings being to attempt to gloss over the more questionable aspects of his involvement with the company. Given the low esteem in which the Board was by now held by the shareholders, such support would be of only limited value in the battle that lay ahead.

The Committee's report revealed a number of serious irregularities in the management of the company's finances whilst Peto had been at the helm. The account books showed not only that the entire 'Consolidated Fund' had been handed over to him without payment, but that he had used this stock to authorise the issue of £390,000 of debentures between August and October 1864. The law stated that debentures could not be issued until at least half the original share capital of a company was fully paid up. The LC&DR directors, when they formally responded to the report, could only say that 'it was considered that loans could be more easily obtained upon paid up stock than upon shares'. A similar irregularity was shown to have occurred with the 'City Lines' debentures, the entire issue of £333,000 having been obtained by Peto, when the only payment made on the ordinary shares was the £300,000 subscribed by the GNR. To make matters even worse, the Committee also discovered that Peto had held on to £128,000 of these debentures; when he should have relinquished them under an agreed procedure, whereby he automatically surrendered each batch as it was sold to the public. The inference was that he had used the outstanding debentures to secure further cash advances or loans on behalf of the LC&DR. The Committee did not suggest, however, that he had appropriated any of the debentures. Although there was implicit criticism of him throughout the report, the company's Secretary, W. E. Johnson, was held most to blame, because he had known of the discrepancy, done nothing about it and endeavoured to conceal the matter from the Committee. Peto must have been relieved to have found a scapegoat, at least for the time being, although the suspicion remained that Johnson had either been in league with him all along, or at least had endeavoured to protect him for as long as possible. Lacking sufficient evidence to charge either man with deliberately over-issuing debentures, the Committee could only record its view that:

> No excuse or palliation can be offered for the very loose and negligent manner in which the debentures have been handled and that the transaction was entirely unjustifiable.[35]

But when it came to the 'Eastern Extension' debentures, the Committee had concrete evidence, upon which to indict both Peto and Johnson. This took the form of letters that had passed between Johnson and Peto's Financial Manager, Charles Christian, in April 1866, just before 'Black Friday'. Extracts were published to show that the two had exchanged 'receipts', each for £429,000, Johnson's proportedly being 'in anticipation of calls upon the contractors' Eastern Extension A and B shares' and Christian's 'in respect of our contract for the construction.' The Committee revealed that:

> These feigned payments were then entered in the books of the company, the only authority for the entries being the receipts. Upon this, statutory declaration was made and the certificate of a Justice of the Peace obtained … to bring the borrowing powers into operation, whereupon the full amount of the debentures authorized (£356,000) was issued and are still outstanding.[36]

Surprisingly, no mention was made in the report that very little actual work had been carried out on the 'Eastern Extension', which could only mean that Peto had needed the debentures to meet the cost of work he had carried out elsewhere. Of course obtaining payment for work before carrying it out, was by no means a new stratagem for Peto. There is no evidence, however, that on the earlier occasions he had obtained debentures on false pretences in this way, although not to say that he had never done so. The revelations in the report put Peto, Johnson, and by implication the whole LC&DR Board, in serious danger of criminal prosecution. Set against the irregularities over the issue of the debentures, the Committee's further allegation that the directors had been unduly lax over the terms of the contract, was of minor consequence, even for Peto.

Peto must have attended the meeting of the LC&DR shareholders, held immediately after the publication of the Committee's report, in a state of considerable unease. It did not turn out to be as traumatic an experience as he had feared however, his case being assisted by being called to account for his actions comparatively late in the proceedings, after most of the shareholders' wrath had been expended upon the directors of the company.

Plate 9a: **The Folkestone Viaduct, designed by William Cubitt, built by Peto.**
(*Illustrated London News* 3/2/1844)

Plate 9b:
The Folkestone Viaduct in 2008. *Author*

Plate 10: **An Eastern Counties train passing the 12th century Ely cathedral and heading south towards the level crossing on the Newmarket Road before entering Ely Station.**
From a watercolour by Edward Duncan c.1850, courtesy of the Elton Collection

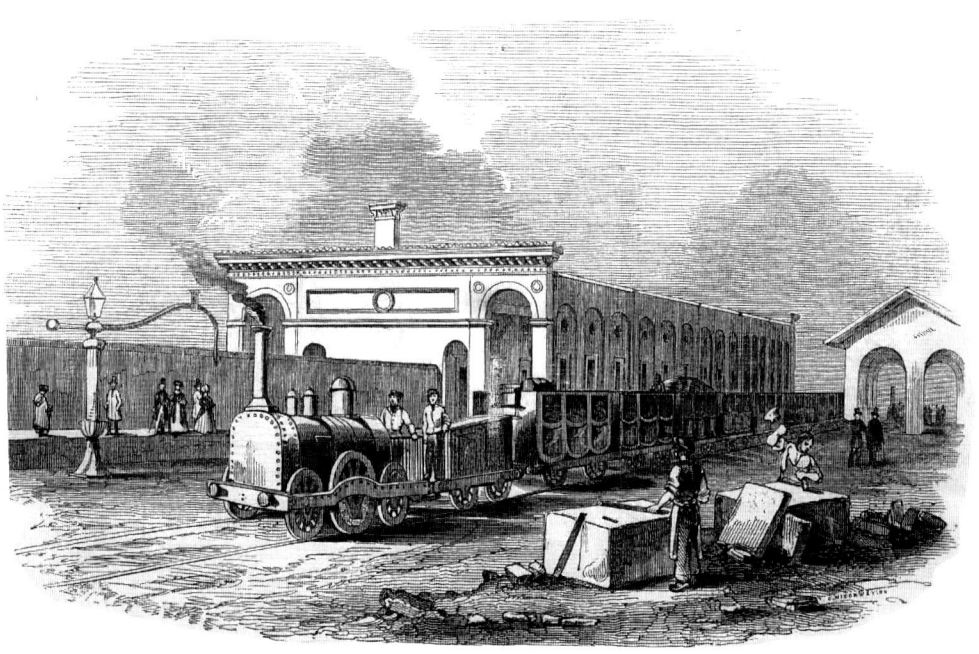

Plate 11a: **Cambridge Station.**
(*Illustrated London News* 2/8/1845)

Plate 11b: **The Royal Victoria Hotel with the station alongside at Colchester
at the time of the extension of the railway to Ipswich in 1846.**
(*Illustrated London News* 20/6/1846)

Plate 12a:
Norwich Station in 1845.
(*Illustrated London News* 2/6/1845)

Plate 12b:
Bust of Peto at Norwich Station.
J. D. Bennett

Plate 12c:
Brandon Station.
(*Illustrated London News* 2/6/1845)

Plate 13a:
Somerleyton Station.
J. D. Bennett

Plate 13b:
**Peto's crest over entrance door
of Somerleyton Station.**
J. D. Bennett

Plate 13c:
**Cottages at Somerleyton – part of
improvements in the village carried out
for Peto by John Thomas.** *Author*

Plate 14a: **East front of Somerleyton Hall.**
(From the 1861 sale catalogue)

Plate 14b: **The west side of Somerleyton Hall today.**
Author

Plate 15a:
Peto as squire of Somerleyton.
(P. Thestrup: *Dampen binder Danmarksammen*, Vol.1)

Plate 15b:
Edward Ladd Betts.
(Institution of Civil Engineers. *Biographical Dictionary of Civil Engineers,* Vol.2)

Plate 15c:
George Parker Bidder.
(P. Jay: *The London, Tilbury & Southend Railway*)

Plate 15d:
James Beatty.
(P. Marsh: *Beatty's Railway*)

Plate 15e:
Thomas Brassey.
(A. Helps: *Life and labours of Mr Brassey*)

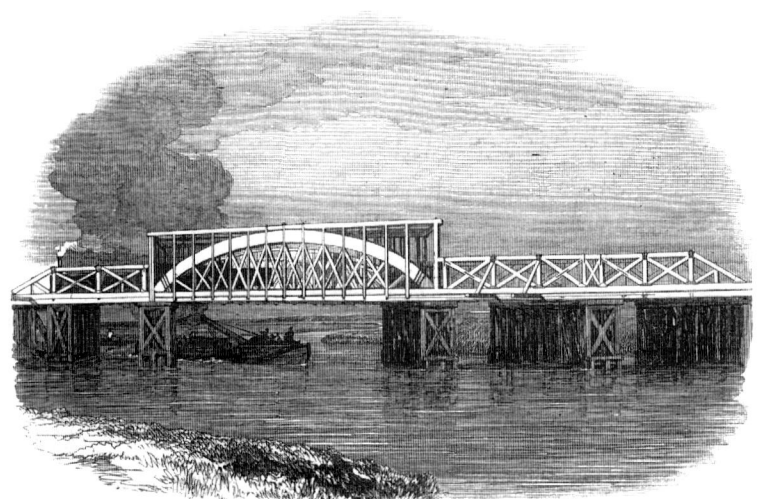

Plate 16a:
Swing bridge at Trowse.
(G. Dow: *The First Railway in Norfolk*)

Plate 16b:
Bardney Bridge on the Great Northern Lincolnshire loop line in the 1840s.
(*Illustrated London News* 28/10/1848)

Plate 16c:
Bardney Station. Some evidence here of excessive facilities for its rural situation.
(*Illustrated London News* 11/11/1848)

Nevertheless Peto did not escape entirely unscathed. When the matter of the outstanding 'City Lines' debentures came up, one shareholder went as far as to say that 'the man who would appropriate £128,000, the money of the company, ought to stand at the bar of the Old Bailey'. His was the sole voice for such extreme action, although there were plenty of groans and some barracking when Peto eventually rose to address the meeting. When this had subsided he proceeded to cannily side-step the question of his excessive 'City Lines' debenture holding, saying the loans he had raised upon them had been entirely a matter between himself and the financial institutions which had advanced money upon them. The ploy worked and he managed to avoid closer questioning on the matter. But when he turned to the 'Eastern Extension' debentures, the weight of evidence was so overwhelming, his only defence was to try to shift the blame on to those who had been involved with him in the irregularities. He told the meeting :

> These debentures were issued – whether rightly or wrongly, it was not for him to say – by the writing up of that stock [i.e. the ordinary shares] against the members of his firm and upon that they were so issued. The writing up had the approval of the solicitors of the company … and nothing irregular was intended on the part of the firm [i.e. Peto, Betts and Crampton].[37]

He declined to comment upon the legality of the whole transaction and was not pressed on the point. There was some vociferous dissent, however, when he went on to declare it was his intention to complete the 'Eastern Extension'.

Then, either in an attempt to attract sympathy or to divert attention from the main points at issue, Peto endeavoured to persuade his audience that he had been a martyr in the company's cause, pointing out the assistance he had rendered the Metropolitan Extensions project, and in particular the the guarantees he had given, which had enabled the loans to be raised that had paid off the Lloyd's bondholders. He said he had renewed these guarantees just before the May 1866 debâcle and claimed this action was the sole cause of his current difficulties:

> It was entirely because of this responsibility which he and his firm undertook, that he appeared that day after thirty years of industrial life, with the solicitor of his creditors on his right hand.[38]

What he failed to add was that he had paid nothing for the scrip upon which those loans had been raised in the first place.

Peto probably thought he was on safer ground when he defended himself at the end of the proceedings against the claim that he had overcharged for the construction of the railway. The rates agreed initially for the work were in his opinion fair and 5% below those paid by the Board of Trade for similar work, a comparison he had made twenty years earlier, when the Eastern Counties company's Committee of Investigation had scrutinised the rates charged for some of its stations. In the case of the South London line, he said he had charged the same prices as the contractor the Brighton company had employed for its portion of that line. He did forget to add that the terms he had originally agreed with the LC&DR for the Metropolitan Extensions had been replaced by the overall fixed-price contract he had imposed upon the company quite early in the construction of the line.

Despite the implicit admission of complicity, it had been a skilful performance and as long as the LC&DR directors stood by him, Peto appeared to have a good chance of avoiding taking most of the blame for what had happened whilst he had been in charge of the company's finances. There still however remained the matter of the share issues he had masterminded on the company's behalf, to say nothing of the complex relationship between his own and the LC&DR's finances, which were potential grounds for contention when it got round to sorting out all its accounts. These were clouds on the horizon, which threatened more serious trouble for Peto in the not too distant future.

Besides conniving at the over-issue of debentures for the 'City Lines', the LC&DR directors in their written response to the Committee of Investigation's report, had supported Peto on every point raised. Regarding the execution of the works they claimed that no contractor could have been found 'of greater power of capital or higher character' and that the terms of his contract had been approved by an independent assessor, although they had to admit that the engineer in question had been Sir William Cubitt, the father of their own engineer, who had been seventy-five years old at the time and was now conveniently dead! It was also claimed that advice had been sought on the financing of the scheme from persons in the City 'competent to form a judgement' on such matters but they were not named, whilst the accompanying submission that the Board contained within its number, men well acquainted with money matters, was not very convincing. The main weakness of the document, however, was its failure to offer any acceptable explanation for the irregularies in the issue of debentures; as far as those for the 'City Lines' were concerned, it was claimed that the contractors were significantly in credit with the company at the time, which was almost impossible either to prove or disprove, as Peto was raising capital at the same time as he was charging for the works. The only reference made to the 'Eastern Extension' issue was an irrevelant statement to the effect that the offending debentures had now been returned by the contractors, as they were unable to complete this line.

On 22 October 1866, only ten days after the shareholders meeting, Peto had the somewhat different but by no means easy task, of defending himself at a meeting of his constituents at Bristol. He had been returned as one of the city's two MPs the previous year[39] and felt obliged to give those who had elected him some explanation of his part in the LC&DR affair, which was already threatening to put an end to his political career. He was sufficiently astute also, to realise that the occasion provided an excellent opportunity to address a wider audience and to vindicate himself in the eyes of the general public. To this end a verbatim report of the proceedings was published shortly afterwards.[40] Craftily, however, this meeting at the Colston Hall, was held at very short notice with the attendance restricted to his political supporters.

The Bristol speech was longwinded and self indulgent. As might have been expected, the arguments Peto put forward were very similar to his earlier statements to the LC&DR shareholders. He maintained that that company and its officials were responsible for whatever irregularies had occurred, quoting as evidence a number of letters that Freshfield & Newman, the LC&DR's solicitors, had written to him. He repeated his claim that the £128,000 of outstanding 'City Lines' debentures were of concern only to himself and the finance companies which had issued the loans, but now added that in his opinion the documents that had been issued had not really been debentures at all and he had been 'perfectly at liberty to redeem them when he pleased.' He used the same ploy to explain away the 'Eastern Extension' debentures, which he told the meeting, had really been 'documents … lodged … with his securities, not put upon the company's register', adding that 'they were not entered in the company's accounts and they did not appear in the accounts published to the shareholders'.

By the time of the Bristol meeting, Peto had probably come round to realising that he was largely on his own and that the LC&DR directors could no longer protect him from what was developing into a two pronged assault from the shareholders against both the Board and himself. So he had no inhibitions in acknowledging that he had taken shares from the company at substantial discounts, pointing out, quite justifiably, that he 'could not take at par that which would not realise par.' The discounts had not been excessive in his submission, considering the amount of financial support he had been giving the LC&DR at the time. He concluded with the bold and uncontestable, if hypothetical statement, that taking everything into account, he and his partners would have been better off if they had taken the works at the lowest competitive price and been paid in cash.

As a public relations exercise, the meeting proved reasonably successful but, realising there was still much to be resolved before he could extract himself from the LC&DR affair, Peto threw out the suggestion at the end of the proceedings that the best way of resolving the issues in contention might be to set up a parliamentary inquiry, adding that he would move a resolution to that effect during the next session. This was a shrewd move because, besides delaying any proceedings in the courts, it ensured him a fairly sympathetic hearing from his fellow MPs, especially those in his own party. When he eventually came to raise the matter in the House, his request was refused, although both Gladstone and Disraeli made sympathetic comments upon his past achievements. By that time his financial predicament had effectively ended his political career and he did not contest the ensuing election in 1868.

The publication of the Committee of Investigation's report generated adverse criticism of both the LC&DR and Peto in the press. Earlier reaction to the events of May 1866 had mostly condemned the excessive speculation of the preceding years and the activities of some of the smaller railway contractors who had suffered the same fate as Peto. The report gave the desired ammunition, however, for an attack upon what was now seen as the worst example of a railway scheme based upon speculative capital. This was spearheaded by *The Economist*, which published a vitriolic article on the subject a few days after the Committee's findings had been made public, titled: 'The history of the London, Chatham & Dover Railway and the lessons to be learned from it'. After branding the scheme 'a contractor's railway … made by borrowing money, not subscribing capital', it proceeded to castigate the methods Peto had used to finance it:

> Since 1860, Peto & Co. have subscribed for the whole share capital of the company of every kind and such capital was subsequently placed upon the market by Messrs. Peto & Co., either on their own account or in some instances, as they allege, on behalf of the company. The loans were often secured by Peto & Co.'s acceptances and have been floating about Lombard Street these six years, growing even larger and larger. A more wasteful mode of getting capital cannot be conceived.[41]

On the matter of the over-issue of debentures for the 'City Lines' the criticism was even more scathing:

> The London, Chatham and Dover did through Peto & Co. pledge all its debentures and when common investors wanted some, issued new ones and trusted to Peto & Co. redeeming the old ones, which they did not always do. New and old ran on together and a large amount of illegal debentures is outstanding still. Every one will agree … that such a policy was unjustifiable in Peto & Co., and unjustifiable in the directors.[42]

The 'receipts' which had been used to authorise the issue of the debentures for the 'Eastern Extension' were described as 'two mutually destructive functions' whereby the contractors had acknowledged 'the receipt of money never paid on account of work never done' and the company 'the receipt of capital subscribed by the contractors, which was never subscribed'.

Both the LC&DR and the contractors were, in the view of *The Economist*, responsible for the disaster, the company for 'rushing to make a railway which the moneyed part of the World were not ready to make', and Peto for behaving as a contractor 'subject to no competition and financial agent subject to no control'. But there was no doubt who was held to be the more responsible.

> The name of Sir Morton Peto has long stood very high for integrity. Perhaps he did not know, at least not in full and exactly how his name was being used, though he was bound to know. But the result is the same. By virtue of Peto's name, the public have been robbed, and they should hereafter be slow to give a similar trust to any other name.[43]

Peto did not respond, preferring to remain silent, as was his way when subjected to personal attacks. At the time it seemed possible that the LC&DR might follow him into insolvency, which would have made the shareholders even more hostile towards him, but with outside assistance and skilful management by a reconstituted Board, the company managed to avoid this. Once the company's finances were on a sound basis, the shareholders began to benefit from the revenue earned by the lines Peto had built for them at such great personal cost. This was small consolation for him of course, although probably those who suffered most from the whole affair were the investors in the fringe banks and discount houses, which had backed the venture through him. As a investor in the bankrupted Imperial Mercantile, Peto lost twice over.

BANKRUPTCY PROCEEDINGS

THE verdict of the press made the LC&DR shareholders even more determined to exact full recompense from Peto for the loss they considered he had caused the company.[44] Working out what was due was not a simple matter, given the complexity of the transactions involved and the close-knit nature of the relationship between the contractors and the company over several years. The directors took the easiest course open to them and claimed a refund based upon the full face value of all the shares Peto had handled, which added up to the colossal sum of £6½million. This was far beyond anything Peto and his partners could possibly pay, and when it became clear that compromise was out of the question, the final settlement was left to the courts. *The Times* predicted a legal contest as protracted as that between McIntosh and the Great Western Railway,[45] which would have reminded Peto of his own good fortune thirty years earlier in being able to settle his dispute with the GWR satisfactorily out of court. His natural inclination would have been for a negotiated settlement but there was much more at stake now than a substantial sum on the construction account. Consequently, along with Betts and Crampton, he was subjected to the indignity of appearing before the Court of Bankruptcy in February 1868. The LC&DR was the only claimant, so presumably he had managed by this time to settle with his other creditors.

After a brief initial hearing, the Court adjourned to allow both sides to gather evidence. When it reconvened two months later, Peto was inevitably the central figure in the proceedings. It is difficult not to have sympathy for his co-defendants; Betts, whose custom appears to have been to give Peto a free hand in financial matters and Crampton, who had, as it would soon be revealed, been kept largely in the dark. Both were to pay dearly for the faith they had had in their senior partner's financial wizardry.

Peto presented his case personally in a confident and business-like manner, hardly needing the assistance of his counsel. He not only insisted, as on earlier occasions, that he was totally innocent of any wrong-doing but strove to prove that he had had the LC&DR's interests at heart at all times and his present predicament was entirely the result of being oversupportive of the company. He began by declaring he had not been in any way involved in the promotion of the company, or its finances, prior to being asked to assist in meeting the cost of the 'Metropolitan Extensions' in 1863, his role up to that time having been simply that of contractor. On the matter of the additional stock 'created' to cover the cost of paying off the Lloyd's bondholders, he had not been obliged to make any payment for this, because the company had agreed earlier to allow credit beyond what was due to him from the works. He claimed he had not been aware it was illegal to combine the new stock created at that time with the unsold Metropolitan A shares, whilst the debentures that had been subsequently issued had been paid for through the works.

Peto turned next to the 'City Lines'. He told the Court that the 'whole of the financial arrangements' here, including raising loans upon the debentures, had been arranged by Mr Johnson but could not deny that he had signed the subscription contract, or that £133,000 of these had been issued to him personally. He insisted, however, that, like the 'Consolidated Fund' debentures, the cost had been covered through the works, although he admitted that only £100,000 had actually been 'paid for' in this way. He said that the disputed debentures had been used to secure a loan of £75,000 from the Union Bank. As for the £128,000 of debentures left outstanding in his name, these represented 'a sum otherwise applied at the request of the company' at a time when he was redeeming a large number of debentures. This was a different line from that he had taken at Bristol, when he had said the sum involved had been a matter between himself and those who advanced him money on these securities.

The issue of the 'Eastern Extension' debentures was the most difficult case to answer. Although Peto sought, as at Bristol, to place overall responsibility for the transaction upon the LC&DR solicitors, Freshfield & Newman, he was obliged to account for his part in it, and in particular the 'receipts' which had been exchanged between Johnson and Christian. At first he tried to pass over the matter but when pressed by Counsel for the company, told the Court that his part of the arrangement had been handled by Christian, as he himself was at home sick at the time and had only learned about it on his return to business a fortnight later. As for the 'receipt' Christian had accepted from Johnson, this had been 'in anticipation of the works'. The phrasing was necessary, as he had been credited in the company's books at the time with £548,000 for work, when the engineer's certificates only entitled him to £70,000. This discrepancy had come to light since the Committee of Investigation's report had been published, and called for an explanation. The best Peto could manage was to say that the construction of the 'Eastern Extension' had been delayed on account of mistakes in the engineer's original plans and the Government's opposition to the proposed route near the Royal Observatory. He was obliged to admit, however, he had found it necessary to suspend work in the Camberwell area for lack of funds.

Although Peto was not interrogated as rigorously as he might have been on any of these contentious matters, the overall impression he gave was of a shifty, behind-the-scenes operator, for whom the ends justified the means, and whose financial manipulations had only been revealed by the crisis he had helped create. In the end he had to admit that the 'Eastern Extension' debentures had been used to finance other parts of the scheme, telling the Court he had exchanged £50,000 of these for provisional bonds issued 'for another line'. He also said he

had needed to replace securities pledged against loans from the Union Bank as 'the old gentleman in charge there', Mr Scrimgour, always wanted his stock 'fresh and fresh'. At this the sombre atmosphere of the court room lapsed into tittering for a few moments.

As at Bristol, the only time when Peto appeared to be on secure ground, was when the matter of the discounting of the LC&DR share issues was broached. Although obliged to admit he had on occasions reaped substantial personal gain from these sales, citing the Metropolitan C stock, which he had received from the company at 75 and later sold to the public at 87½, he claimed these gains were necessary to offset his losses, adding that the contractors had been 'the conduit pipe for the disposal of most of the A and B shares'.

After two days the proceedings were again adjourned, the Court not reconvening until the end of May, when it received a statement on the contractors' own accounts, which, it had been alleged earlier had been in a state of some disarray, although this was subsequently denied. A thorough scrutiny of the books now showed that the LC&DR had been charged £700,000 more than it should have been on the construction account, disproving an earlier statement of Peto's that the company owed him money overall for the works. It was also revealed that some money which had been credited jointly to Peto, Betts and Crampton, had been transferred to another account only in the names of Peto and Betts, evidence that Peto was using LC&DR funds to support other activities. Crampton remained loyal to him when questioned on the matter, however, saying that, although he had been unaware that this had happened, he was quite happy for Peto to make use of their joint earnings, as he had done earlier on an overseas contract they had carried out together.

A six-week recess then followed, during which an agreement was reached in private between the parties, preventing any further scrutiny in public of matters which were proving increasingly delicate for both sides. In the end the only realistic solution from the company's point of view, was to get the maximum possible recompense from Peto and his colleagues, which had to be very much less than the £6½million originally claimed. The terms of the final settlement, formally presented to the Court in July, provided for the LC&DR to receive £484,000 in respect of work that had been paid for but not done; of this £365,000 was to be charged to all three and the remainder to Peto and Betts, plus a hefty £800,000 for the shares that had been handed over by the company and were still outstanding.

After the necessary formalities, during which the counsel for the company, claimed that the settlement, although much less than the LC&DR had the right to claim, had nevertheless 'considerably impoverished the estates' of all three defendants. Before the discharges were issued, the Commissioner in charge of the Court, to the relief no doubt of both the recent adversaries, declined to pass any opinion on the legality of the contentious debenture issues, which was where the matter was allowed to rest. It is not known how much of the total bill of more than £1¼million fell upon Peto, but as the senior partner it would almost certainly have been the largest part. He nevertheless had some grounds for satisfaction at the outcome of the proceedings. Besides forcing the LC&DR to abandon its previous claim, which had been unreasonable anyway in the light of the low standing of the company's shares on the market, the threat of a criminal prosecution, hanging over him since the Committee of Investigation's report, had not materialised.

Although much of what Peto did whilst effectively in control of the finances of the LC&DR can rightly be criticised, the shareholders who subsequently viewed his involvement with the company in a totally negative light, did him a considerable injustice. Without his support it seems very unlikely that the 'Metropolitan Extensions' project, even in a scaled-down version, could have been carried through to completion at the time, or possibly ever. Even if his motives were not at any time altruistic – and that could hardly be expected of any contractor – there can be little doubt that he committed all his personal resources and prestige to sustain the scheme. At one stage during the bankruptcy proceedings he had asserted just that, telling the opposing barrister

he had always 'done his best for the company', only to invite the retort 'And for yourself also?' Left for a moment speechless, Peto might well have reflected that the tragedy of the whole affair as far as he was concerned, had been his inability to do either.

Besides costing him a large sum of money, the LC&DR affair had shattered Peto's reputation in the business world and undermined his position as one of the country's foremost railway contractors. The esteem of others was of paramount concern to him, as he openly confessed at the Bristol meeting, when he said that he did not regret either the forfeiture of his position in the mercantile world or the loss of money, half as much as the damage that had been done to his reputation. It had been the basis of his creditworthiness and sustained his self-confidence over many years. This disastrous contract ensured that he would never be the same man again.

A postscript to the bankruptcy proceedings and another embarrassment for Peto, was to have to face scrutiny of his conduct by his fellow Baptists. The Bloomsbury Central Chapel, of which he was the founding figure and most prominent member, set up a committee of enquiry into the whole affair. When they questioned Peto over the financing of the scheme, he had to admit that he had known that some of the subscribers to the scheme, like clerks in his own firm, did not have the resources to cover their nominal investments; but claimed there had been no concealment, as the procedure was 'merely a nominal and legal compliance', an open admission that he had flouted the law. The Baptists accepted this however and in their report declared that they were strongly of the opinion that Peto had not been guilty of any fraudulent or deceitful conduct, or of bribing any of the other parties involved in the case, such as the LC&DR solicitors, as had apparently been suggested in some quarters. He was nevertheless censured for accepting more liabilities than the £2million of assets he told them he had when he embarked upon the contract. If this figure was correct and his liabilities other than to the LC&DR were minimal, he would appear to have been far from totally ruined by the LC&DR affair.[46]

SEVEN

Aftermath
1866–1875

THE CRYSTAL PALACE &
SOUTH LONDON RAILWAY

ALTHOUGH Peto had been forced to abandon the construction of the Eastern Extension, he had managed to complete another project closely associated with the LC&DR before the May 1866 crisis. This was the Crystal Palace & South London Railway (CP&SLR), which had been promoted as a nominally independent line, running from the 'Eastern Extension' at Peckham Rye to a High Level terminus at the Crystal Palace (plate 19a). This was closer to Paxton's resited building than the station Peto had built earlier for the West End of London and Crystal Palace company. The contract for the nine mile long CP&SLR had been placed with him in 1862, for the fixed sum of £756,000, with the provision that he could charge for any additional works on the same schedule of prices as applied to the LC&DR. Although far less expensive than the Metropolitan Extensions, the cost exceeded £80,000 per mile, which was hard to justify on construction grounds alone, because, although there were two tunnels along the route, the CP&SLR crossed a still comparatively thinly populated district. The expense rather reflected the means by which it was financed and the degree of support it needed from its contractor.

The capital authorised for the CP&SLR had been £900,000, of which £225,000 was to be raised through loans, Peto most likely being involved with these. He also undertook in 1863 to guarantee the interest on the company's 'B' stock, in return being handed the shares at 5% discount. Even more significantly, the directors had agreed

to hand him £100,000 in debentures in advance of works being completed. Fortunately he finished the line in 1865, preventing the company from being dragged into the financial morass along with the LC&DR, although it did not escape entirely. When the CP&SLR directors tried to obtain compensation for Peto reneguing on the share guarantee, they discovered there had never been any binding agreement over the payment of the 5%. They were also unable to pursue a claim against him for defective rails. But the most serious long-term consequence, as with too many of the companies that Peto had supported, was the railway's gross overcapitalisation, which left it with an excessively heavy burden of interest charges. To make matters even worse, the railway proved something of a white elephant from the start, never attracting the traffic it was designed to handle, which was convincing proof long before the decline in its popularity, that the Crystal Palace at Sydenham did not need a second rail link with central London.

THE METROPOLITAN
DISTRICT RAILWAY

PETO had another contract in hand in London in 1866 besides the LC&DR. This involved the construction of five miles of underground railway from Westminster to Cannon Street, which he had undertaken the previous year, jointly with two other leading contractors, John Kelk and Waring Brothers, for the Metropolitan District Railway (MDR). Peto had been involved through his contacts with the Oxford, Worcester & Wolverhampton solicitor,

John Parsons, ten years earlier in an abortive attempt to launch the Metropolitan Railway (MR), the first underground railway project in London. This had been left in the end to others, including Kelk, to construct.

The new scheme was the first stage of an ambitious project to extend the MR from Paddington at one end and Farringdon Street at the other, and thereby create what would eventually become the Circle line. The MDR was floated as a separate concern in the early 'sixties and like its predecessor, would be built on the 'cut-and-cover' principle, technically less demanding than deep tunnelling through the London clay. The scheme had been given a fillip when Parliament approved the proposal of the Metropolitan Board of Works (MBW) to lay one of its new main outflow sewers along the north bank of the Thames, parallel with the proposed railway, and the two projects were carried out in unison. Peto had for long been an advocate of sanitary reform and strongly supported the scheme when it was debated in the House of Commons. He might have been expected to do this, as his firm had been involved in the laying of some of the earliest brick sewers in the capital. At a more modest level, as long ago as 1852, he had been involved with Henry Cole, in the establishment of the first public convenience in London.

Despite being carried out jointly with a public body, the MDR still proved an expensive undertaking, the price for the initial five-mile section after considerable haggling being agreed at £1,710,000. This made it more expensive per mile than any of the lines Peto built for the LC&DR, except the 'City Lines'. Whilst the nature of its construction accounted for a large part of the cost, as with the Metropolitan Extensions, the company relied upon its contractors for a considerable proportion of its capital and was obliged to compensate them accordingly.

At the beginning the MDR directors had hoped to have the best of both worlds, contractor support and economical construction. Their minute book records that one of the first instructions to their engineer, John Fowler was:

To make such contract for the construction of the works … as, whilst it shall secure the

company the best means of obtaining their capital shall … be in his judgement fair and equitable in respect of prices and conditions, without pledging the company to proceed beyond the capital for the time being subscribed.[47]

Although Peto had worked with Fowler on two earlier occasions, it seems unlikely in this instance that he influenced the engineer's appointment, as he was the obvious choice for the job, having carried out the construction of the Metropolitan line. As for the contractors, Kelk's previous experience was obviously invaluable, whilst the expertise Peto's firm had acquired in laying brick sewers over a number of years, could be put to good use and would be an asset when it came to negotiations with the MBW chief engineer, William Bazalgette.

The MDR directors soon found they had to fall in line with the contractors' requirements, as from the start, the powerful triumvirate was to raise most of the capital needed for the project. There were similiarities with the LC&DR: the contract was drawn up after private negotiations with the contractors, without any other tenders being sought and it was agreed at an early stage that they should be paid £45,000 per month above 'that required for the works.' Another ominous development was the directors' acquiescence in April 1865, to the request that they should deposit £200,000 with the Imperial Mercantile Credit Association, the fringe bank with which Peto had close links. What use was made of this large sum of money in the short term is not known but it could have helped Peto indirectly at least in overcoming the problem of raising cash needed to complete the 'City Lines'.

Financial constraints had obliged the MDR to limit construction to the section of the line between Westminster and Cannon Street, although at one stage the directors had asked the contractors for help to extend eastwards from Cannon Street to Tower Hill. Although it is not known how the contractors divided the work between themselves, it seems quite likely that Peto hoped to build a large part, if not all, of the Tower Hill line. When disaster struck in May 1866, he appears not to have been greatly involved in the construction of the line in hand; having limited himself to some minor works at Blackfriars, which would have fitted in with his LC&DR undertaking. He probably hoped to transfer a significant portion of his workforce onto the MDR once the LC&DR was finished. Certainly neither the MDR, nor, as far as can be ascertained, either of his partners made any claims upon Peto's estate at the time of his bankruptcy, although there must have been some private settlement of accounts. His place on the consortium was taken by Lucas Brothers.

THE CORNWALL MINERALS RAILWAY

WITHIN a couple of years of returning to England from Hungary (see p.109), Peto took the first steps to relaunch himself as a railway contractor. He knew, however, that he would have to execute any new work without the assistance of either of his old partners; Brassey had died in 1870 and Betts had retired in poor health to Egypt. His next and final contract would be the only

one he completely executed on his own. This must have added to the strain for a man now in his early sixties, bearing the scars of his recent misfortune, and whose mental faculties were probably past their peak.

Unwisely, Peto again chose to become involved with a highly speculative venture. The Cornwall Minerals Railway (CMR) was the brain-child of the London financier W. R. Roebuck, who had ambitious plans in the early 'seventies to exploit the mineral resources of Cornwall, in particular iron ore, which was in short supply at the time. Roebuck floated a company with the grand title of The Cornish Consolidated Iron Mines Corporation (CCIMC) to extract the ore, and the CMR was promoted to convey it and other minerals to the coast for shipment to South Wales and the North of England for smelting. Peto invested in the CCIMC, as well as the CMR, the contract for the latter being his reward.

Authorised in 1873, the CMR incorporated a number of existing mineral lines, including the system of horse-drawn tramways that had been laid down thirty years before by a local landowner, J. T. Treffry. These were now to be

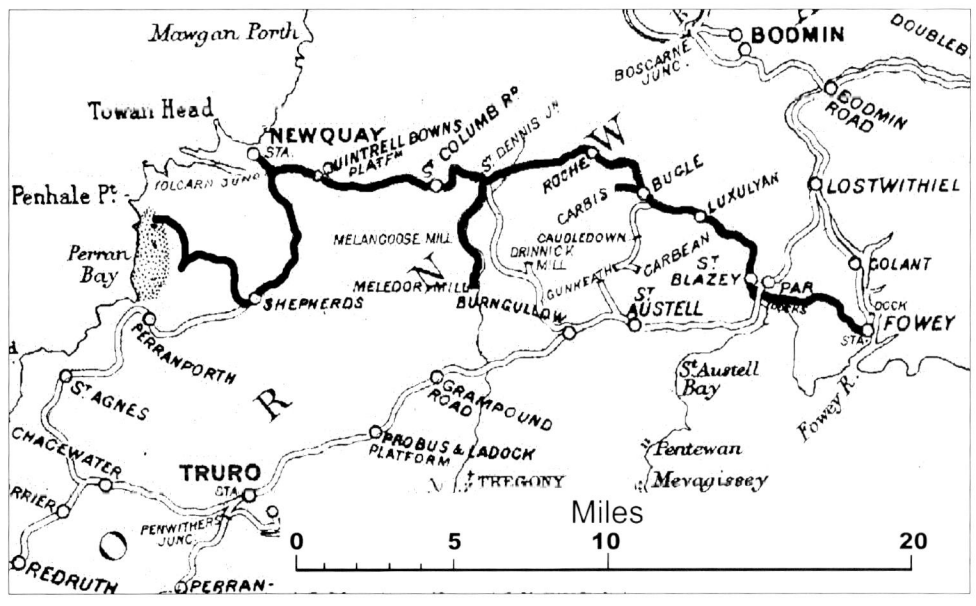

Fig: 7: **The Cornwall Minerals Railway built by Peto in the 1870s.**

upgraded for locomotive working. An additional twenty-six miles of line were also to be constructed, including a deviation to avoid a steeply-graded section of the tramway, a line from Par to the deep-water harbour at Fowey and branches to a number of mines. Peto took an active part in preparing the plans, touring the district with Roebuck in 1872, but failed to persuade the landowners around Fowey to accept the cheapest route to the harbour.

Peto had subscribed £137,500 when the scheme was launched, which ensured he had no rivals for the contract, although, probably recalling what had happened to the LC&DR, the CMR directors stipulated he must pay the sums due on the 'calls' in cash. Although the scheme had been launched on the crest of a new boom in railway investment, the company soon found itself acutely short of funds, requiring Peto to accept additional shares as payment for the works. This put considerable strain upon his limited reserves of capital, which he could not so easily supplement with loans as in earlier years. He was obliged quite early on to inform the CMR directors that he might not be able to complete the railway, unless they assisted him in obtaining additional funds. These were not forthcoming, however, and by the time the railway was nearing completion, Peto had reached the end of his tether, being unable early in 1874 to pay the bills for some of his materials. There was certainly the serious threat of a second bankruptcy, with the added indignity of having the works re-possessed by the company. But in the end the CMR did provide some limited assistance, agreeing to apply some of the balances Peto had accrued from the works, to meet the debts with his suppliers.

Although Peto managed to complete the CMR contract, it would appear to have been at a considerable personal cost. He gave a very frank account of his financial predicament to William Pease, the steward of the Boconnoc estate in Cornwall, who recorded in his diary:

Sir Morton Peto came this day and gave me particulars as to the state of his affairs. The Minerals Railway contract has ruined him. He has not a shilling of his own and owes thousands of pounds to those who have lent him money to enable him to proceed with the contract. The railway company has taken all his plant and he owes a large sum for iron rail. He does not know what course his creditors will take, but there is nothing more for them. In order to make the payments which has [sic] been made, his plate has been sold and also his wife's jewels. He says the estimates for his contract were too little in consequence of the great rise in wages and the price of material.[48]

It is not known why Peto made this admission, but it was probably connected with land he had acquired from the estate to build the railway. Whatever the precise circumstances, it must have been an exceedingly humiliating experience for the former owner of the Somerleyton estate to have to plead poverty to the steward of Boconnoc. In addition to whatever loss he made on the contract, the failure of the CCIMC in 1875 and the insolvency of the CMR itself the following year, certainly compounded his problems.

With the possible exception of Hanwell Viaduct, the CMR had probably been the only British contract where Peto lost money on the works themselves. He had clearly under-estimated his costs and made insufficient allowance for the inflationary effect of the early 'seventies boom. He may have got himself into this situation through an overzealous desire to assist the company, as there is no evidence that W. R. Thomas, the CMR engineer, had demanded a particularly low price for the works. These moreover, appear to have been charged for on a schedule of prices, but this did not compensate for having been agreed at the end of a recession, when the line was built during the succeeding boom. As a result, Peto's attempted come-back failed and his forty-year career as a railway contractor ended on a decidedly downbeat note. Although he had lost appreciably less than in 1866, in his reduced circumstances the sums involved would have meant appreciably more, pound for pound, than on the earlier occasion.

EIGHT

Overseas Ventures

ETO began his work overseas in 1851, when he was proceeding slowly on the Oxford and Birmingham line and the post-Mania depression still made it difficult to obtain new contracts at home. That year, however, he began the collaboration with Thomas Brassey, which would last until the end of the 'fifties, and it was this which gave him his first taste of work overseas. Brassey had pioneered British contracting on the Continent a decade earlier, when he built the railway from Paris to Rouen, and later extended this to Le Havre. Finding himself without a partner after the death of William McKenzie, he was glad to have Peto's support, when he obtained another important contract in France. This was to construct the sixty-seven mile line from Lyons to Avignon, which formed part of the trunk route linking Paris with Marseille. Although they successfully completed this in 1852, the growth of protectionism that followed Napoleon III's coup d'état and the establishment of the Second Empire that year, together with the growth of French railway experience, made it increasingly hard for British firms to obtain new contracts in the country.

THE ROYAL NORWEGIAN RAILWAY

ETO had in the meantime obtained a contract in Norway. Although only forty miles long, the Royal Norwegian Railway (RNR) was the first line to be built in the country. The scheme owed much to the efforts of the British Consul in the capital Christiana (the modern Oslo), John Crow, who not only persuaded the Norwegian government to lend its support but also managed to interest a leading London financier, Lewis Ricardo, in the venture. Peto appears only to have become involved at a comparatively late stage, and probably was introduced to Ricardo by Robert Stephenson, who had carried out a survey of the route for Crow. Stephenson's plans were for a line running inland from the capital to Erdsvold, to link up with the Mjøsa Navigation, which provided access to a large tract of the forested interior. The railway promised to open up new sources of timber, which may have been enough in itself to attract Peto, as he made extensive use of this himself and it was in short supply at home. In addition, the geographical situation of Norway made it well placed to be linked by steamship with Lowestoft, the port on the other side of the North Sea, in whose future he had a substantial personal stake.

Under the terms of their contract, Peto, Betts and Brassey, along with Ricardo, undertook to raise half of the company's total capital of £500,000, whilst the Norwegian government agreed to provide the remainder and pay 5½% interest on their overseas backers' investment. This made the venture less risky for the contractors than it would have been had they gone ahead on their own. There was a clause in the contract, however, which obliged them to employ local labour, which may not have been entirely to their liking, as Brassey had used British navvies on at least some of his French works, whilst Peto probably had labourers surplus to requirements back in England. In the end the whole line was subcontracted anyway and executed by George Merrett and Thomas

Earle, both of whom had been involved with Peto in the construction eight years earlier of the Yarmouth & Norwich Railway. Another recruit from those days was George Bidder, who acted as Stephenson's deputy, besides being the Resident Engineer.

Peto crossed the North Sea to represent the contractors at the ceremony of cutting the first sod, which took place at Christiana in October 1851. This was attended by local dignatories, as well as Crow, who gave the project the British government's official approval. The star performer on the day, however, was Peto, who treated the assembled body to a lengthy speech, in which he displayed all the diplomatic skills required of a business man abroad, telling his audience:

> The undertaking so auspiciously commenced today, is fraught with benefit to your country and to my own; it will in the completion, place your fatherland a century in advance; it will develop your natural resource and bring our nations into close contact; and though as capitalists we necessarily look to self interest as the first motive to action, we are not insensitive that there are other motives than mere money value. Our countries have a common sympathy; you have a Sovereign whom you affectionately regard, who not only commands your loyalty but reigns through your love; you have a constitutional country, whose interests are entwined round your sympathies … we too have a Queen who reigns paramount in our devotion and regard; we too have a country and constitution whose well-being is our constant care and we cannot forget that there are historical incidents which bind us closely; and the cordial reciprocity of today makes me feel that the sentiments of your hearts are to unity with mine and that were they audibly expressed, your response would be, were we not Norwegians, we would be Englishmen.[49]

By the time Peto returned three years later for the opening ceremony, the atmosphere had changed. A considerable amount of ill feeling had developed in Norway to the English interests that had masterminded the scheme. In order to carry the railway through fairly difficult terrain and keep down construction costs, Stephenson had been obliged to construct the RNR with steep gradients and numerous sharp curves, which in due course made the railway both difficult and expensive to operate. The Norwegians nevertheless considered with some justification that, at more than £12,000 per mile, they had got poor value for money, especially as Bidder had substituted local timber for iron wherever possible and failed to provide adequate facilities for handling locomotives, or enough rolling stock. These economies could not be attributed entirely to the reduction of £50,000 in the contract price that Peto had been forced to accept, when the company became short of funds. The feeling of many in Norway was that despite his fine words at the inauguration of the project, he had endeavoured throughout to maximise the return on his contract.

Another bone of contention with the Norwegian authorities, concerned the five-year lease the contractors had agreed to take on the RNR. Although they had shipped over locomotives and rolling stock from England, the services proved unreliable and there was also a suggestion that the line was being run down over the period of the lease. Peto and Brassey soon had to face similar criticism back in England from the shareholders of the London, Tilbury & Southend Railway (LT&SR). In the end the Norwegian government lodged a formal protest and called in Robert Stephenson to adjudicate the claim. After considerable deliberation Stephenson eventually managed to arrive at a compromise that was acceptable to both sides. It was on the voyage home from Norway in 1859 after securing final approval from the authorities there for his proposals, that he was taken fatally ill.

Peto probably hoped for more from the RNR than the profits from the contract. During the course of his speech in 1851, he had been eloquent in his praise of the scenic delights of the country, telling the Norwegians :

> Your splendid mountain scenery, your lakes, your unique fiords, your stupendous

waterfalls, will bring all my pleasure-seeking countrymen, to whom the other parts of the Continent are a beaten track, to your shores.[50]

Despite the failure of the hotel at Colchester, Peto continued to interest himself in the idea of catering for the leisure needs of the Victorian middle class at home and may well have envisaged providing the necessary facilities abroad as well. He certainly did acquire some land in Norway at this time, but there is no evidence he proposed to develop this in any way, or even make use of it personally. This was just as well because, as with too many of his ideas, it was well ahead of its time, nearly half a century passing before Norway began to attract British tourists in any considerable numbers.

In the same speech Peto had proposed the establishment of a regular weekly steamer service between England and Norway, which would, he claimed, put Christiana and London within sixty hours travelling time of one another; no doubt hoping that this would be provided by his own North of Europe Steam Navigation Company, working from its Lowestoft base. Nothing came of this, nor were there to be any more railway contracts for him or Brassey in Norway.

THE ROYAL DANISH RAILWAY

I N 1853, before the work in Norway was quite finished, the same contracting team, again with Ricardo, embarked upon another railway project in neighbouring Denmark. Although known from the start as the Royal Danish Railway (RDR), unlike its counterpart to the north, this was not quite the first line to be built in the country; nor did it receive any financial backing from the state, nearly all its capital having to be raised on the London stock market. The contractors, as in Norway, also agreed to lease the line in this case for fourteen years, guaranteeing the investors a return of 6% upon their capital, plus half the profits, the same terms as had been agreed the previous year with the LT&SR, although the RDR lease ran for only just over half the time. The capital needed for the Danish work was only slightly more than that for the Norwegian line, but twice as much had to be raised on the open market. The contract price for the RDR, inclusive of all 'plant, stations and every contingency' was £540,000. As with the LT&SR, the lease represented a substantial long-term liability for Peto and his colleagues, and had the higher element of risk associated with an overseas investment.

Although the Danish government had not backed the scheme financially, it was of considerable strategic significance, as the line crossed the Duchy of Schleswig, which with the neighbouring Duchy of Holstein constituted the base of the Jutland peninsula, and formed a buffer zone between Denmark and the German states, dominated by Prussia, to the south. Although the duchies were nominally independent, they were ruled over at this time by the King of Denmark, Frederick VII, and it was he, rather than the Danish state, who conferred the title 'Royal' on the railway company. The railway was seventy-two miles in total length; the main line ran across the neck of land separating the North Sea from the Baltic, from the small port of Tonning in the west, to Flensburg, an important deep-water harbour; there was also a branch from Husum to Rendsburg to link with the Altona Kiel Railway, which ran south towards Hamburg *(plate 20a)*.

The RDR employed the same engineers, as well as the same contractors, as its Norwegian predecessor. Robert Stephenson was again in overall charge, but in this case appears to have done little besides the initial survey, leaving the entire execution of the line to Bidder. It was an easier railway to construct than the line in Norway, as it passed through a country of low hills interspersed with lush meadows, which must have reminded some of those who crossed the North Sea to build it, of the terrain they had become familiar with in East Anglia. The approach to Tonning by rail today is very reminiscent of crossing the Norfolk Broads, on the line from Norwich to Yarmouth.

The RDR was opened by Frederick VII in October 1854, amidst considerable rejoicing.

There were gala processions in a number of towns along the route, whilst Flensburg celebrated the event with a fireworks display and a royal ball. Peto crossed from Lowestoft to Tonning in *Cygnus*, the flagship of his North of Europe Steam Navigation Company, which he commandeered as his personal yacht for the occasion. After escorting Frederick on a trip along the railway, he entertained the King on board the ship, presiding over the ceremonies dressed in his Deputy Lieutenant of Suffolk uniform. In a private ceremony in the saloon after dinner, Frederick presented Peto with the Order of the Dannebrog, which he said was the 'formal memorial of approval for a great Danish work … executed by a great British capitalist and contractor', as well as 'a souvenir of his regard for an amiable, liberal-minded and accomplished English private gentleman.'[51] After the formalities had been concluded, Peto sailed back to Lowestoft in company with Stephenson and Bidder, whilst the King proceeded on an extended tour of the Duchies by train, the royal carriage having been specially built on Peto's orders, by the Stratford workshops of the Eastern Counties Railway *(plate 20b)*.

The RDR remained a private company for ten years, the lease having four years to run when Schleswig was annexed by Prussia, following her defeat of Denmark in 1864. With deference to the British capital that had been invested in it, the RDR did not, however, become war booty. The shareholders redeemed their capital when the company was purchased by a Prussian bank shortly after the war. This also relieved the contractors of their leasing obligations. The railway was in due course incorporated into the Prussian state system, although the German railway engineers subsequently built a more direct line, running northwards along the coast from the Rendsberg branch. The original line from Flensburg to Husum is now abandoned but the remaining portion to Tonning survives, the tracks still running to the 'new harbour' constructed in connection with the railway *(plate 20c)*. A century and a half after this was built, it is in a bad state; the rails rusting and the only signs of activity, besides the busy nearby road bridge over the river Eider, are a few parked lorries. At Flensburg, however, the extension to the harbour is still in occasional use for goods traffic, although the original station, erected by Bidder and known after the Prussian annexation in 1864, as 'Der Englische Bahnhof', has been demolished to make way for the town's main bus terminus.

Peto had hoped that the RDR would enable him to establish a new line of communication between London and Copenhagen, using the railways he controlled on both sides of the North Sea and his own steamships between Lowestoft and Tonning. The additional traffic this promised to bring onto the Danish line would obviously benefit him and his partners as lessees. Unfortunately he miscalculated the potential traffic and the drawback of a twenty-four hour and often longer, passage from Lowestoft to Tonning. On one such crossing, Peto himself had the unpleasant experience of being stormbound for several days. Although the steamships did convey some passengers and there were even a few excursions laid on for the more intrepid tourists, it soon became apparent that revenue from this source, would be insufficient to sustain the service. The only alternative was freight but, although faster and more reliable than sailing vessels, the company's steamers were expensive to operate and could not compete for general cargoes on the North Sea crossing.

Peto then struck upon a bold plan to provide the necessary revenue for the North of Europe Steam Navigation Company. This was to use facilities provided by the steamships and the railways on either side of the North Sea to transport cattle from the lush pastures of Schleswig Holstein 'on the hoof' to England, where there was an insatiable demand for fresh meat from the rapidly expanding urban population. Although live cattle had been exported from Tonning mainly to the English East Coast ports for a number of years previously, Peto expanded the trade dramatically by employing larger and speedier vessels, which ensured the animals arrived in better condition than if they had been subjected to a long sea passage. Attractive rates were offered for the crossing and Peto used his influence with a number of railway companies to obtain concessions for the transport of cattle on the last leg of their journey. Despite all his efforts the trade did not prosper, however, being adversely affected by the increasing imports of cheap store cattle from Ireland. An outbreak of cattle plague on the Continent delivered the final deathblow. The ultimate misfortune for Peto was the subsequent bankruptcy of the North of Europe Steam Navigation Company *(see p.29)*.

Under the terms of the original concession, the RDR had been granted the exclusive rights for all railway development in South Schleswig but as early as 1853 Peto had set his sights on a much more ambitious concept – for a system of lines running northwards from Flensburg as far as Frederikshavn on the northern tip of Jutland, where a ferry to Gothenburg would give access to the railway that was under construction to Stockholm. This grand Nordic strategy would not only benefit the RDR and his steamship company but also, he thought at the time, further expand the livestock trade with England. As a first step he persuaded the RDR to upgrade the Tonning–Flensburg line with some new bridges and to rename itself the South Schleswig Railway, as a North Schleswig Railway was now projected from Flensburg to the Jutland border. Like the earlier project this was dependent upon British capital and ten years would pass before it was finally built.

The Danish government was more concerned with the political and strategic, rather than the commercial implications of railway develop-ment, so the route to be followed by any line built through Jutland became its primary concern. The military authorities insisted that it was built close to the Baltic coast and Peto encountered formidable opposition when he proposed an inland line, which would better serve the cattle trade. Before long the matter became politically charged and the subject of an acrimonious debate in the Danish Parliament, when strong feelings were expressed against foreign interests tampering excessively in the development of the nation's railway system. But after an abortive attempt to finance a railway on their own, the Danes were obliged to accept the inevitable and Home Minister Krieger re-opened negotiations with Peto in the late 'fifties. He journeyed to Copenhagen in December 1860 to settle the matter with the Danish authorities. By this time his personal objectives had changed considerably. With both the cattle venture and the steamship link no longer viable, his main aim was to obtain the potentially most profitable new contract for building the railway.

Peto summarised his proposals in a letter he wrote to Krieger immediately after their meeting.

Copenhagen 13th December, 1860
To His Excellency,
The Minister of the Interior of the
Kingdom of Denmark

In accordance with Your Excellency's wish, expressed at our interview of this morning, I beg to lay before you my views as to the best mode of carrying out the system of railways under consideration. Of the three modes suggested, I very much prefer that of state bonds and for the simple reason of its producing a better result than any other mode of using the credit of the state, thus giving you the system of railways at a smaller cost than any other system. The system of bonds will involve a leasing of the line. This may be done, as already suggested, the leasing party finding the rolling stock which will be a good and full security …

I would strongly urge on Your Excellency the agreement for the entire system, including the North Schleswig line under one contract. The arrangement for carrying out the work will be very much facilitated thereby, and it would otherwise be necessary to add a not inconsiderable percentage to my estimate. Any uncertainty also, beyond that arising from risk of unavoidable, natural or political occurrences, would seriously interfere with the formation of a general working company for the whole system, an object which I look upon as of the greatest importance when considered in a commercial or military point of view, or with reference to the accommodation of the public.

I have had the opportunity this morning of stating to your Excellency verbally my views on the benefit of unity of working and this arrangement will insure that result from Rendsborg [sic] to Aalborg, and across Fynen to Korsor , connecting across the Belts by a steam-ferry taking all the train of carriages over at once, or a goods train of twenty-four waggons.

This unity of working and under a lease, in which the mutuality of interest of your country and the lessees will be the same, will, I am sure, produce the best result in efficiency, economy and dividends to all parties interested …

I have the honour to remain Your Excellency's very devoted and faithful servant,

S. Morton Peto [52]

The alternative methods of financing Peto argued against, had been a subvention by the state, which he claimed would have to be significantly discounted to raise sufficient cash, and the issue of shares with government guarantees that would also have to be discounted and might not produce enough funds for the branch lines.

Two days later Peto wrote to Krieger again, proposing a sliding scale for the division of the profits of the railway, with the contractors/lessees getting up to 75% if these were low. He also made it clear that the profits were to be net and not gross, having learned something from his recent experience at home with the LT&SR lease.

For Peto a substantial slice of gilt-edged securities, was sufficiently attractive to offset the liabilities of the lease. He managed to carry Krieger along with him but others were less easily persuaded and another heated debate took place in the Danish Parliament. As a result of this, he was obliged to accept a majority of Danes on the Board of the railway company and in the end was only allowed to construct the lines in Jutland. These added up to 270 miles of track, including the line along the east coast from the Schleswig border as far as Aarhus, which he had previously opposed (plate 21b).

The works were carried out in the early 'sixties, whilst the North Schleswig Railway remained a separate concern, despite his plea that it should be incorporated into the larger government backed scheme. In 1863 Peto undertook to build the North Schleswig main line from Vamdrup, on the Jutland border, to Flensburg, together with the branches from Rødekro to Abenrą and Tinglev to Tønder, for £630,000. Like the South Schleswig line it was taken over by the Prussians after the 1864 war.

Although some of the Jutland lines had involved heavier works than those in the Duchies to the south, the terrain of boulder clay and sandy glacial moraine, did not present appreciably more difficulties for Peto than their counterparts in East Anglia. What he did have to contend with, however, was the close supervision of the works by engineers appointed for this purpose by the Danish government. All graduates of Copenhagen Polytechnic, in the view of his agent, F. J. Rowan, they adopted an excessively theoretical approach. Brassey made similar comments, going on to attack the undue influence of the public authorities in Denmark, who even had the power to determine the rates of pay of the workmen parish by parish. He did however praise the Danish workers themselves, who he said were for the most part steady hardworking men and contrasted them with some Swedes he employed for a while, who, like a number of the English workers brought over, had fallen victims to the temptations of the cheap local rye whisky.

The lease for the Jutland lines was still running in 1866, when Peto was obliged to hand over his Danish interests to Brassey. How much profit either made is, as so often, uncertain. Although the charges made for the works were higher than those for the RDR, there was less opportunity to profit from the shares. Nor do there appear to have been any opportunities to cash in on additional works, or indulge in remunerative private speculations. The cattle fiasco had probably put Peto off the latter by this time anyway. He had, however, earlier invested in a company set up to supply gas to much of Copenhagen, Bidder being the engineer responsible for that project.

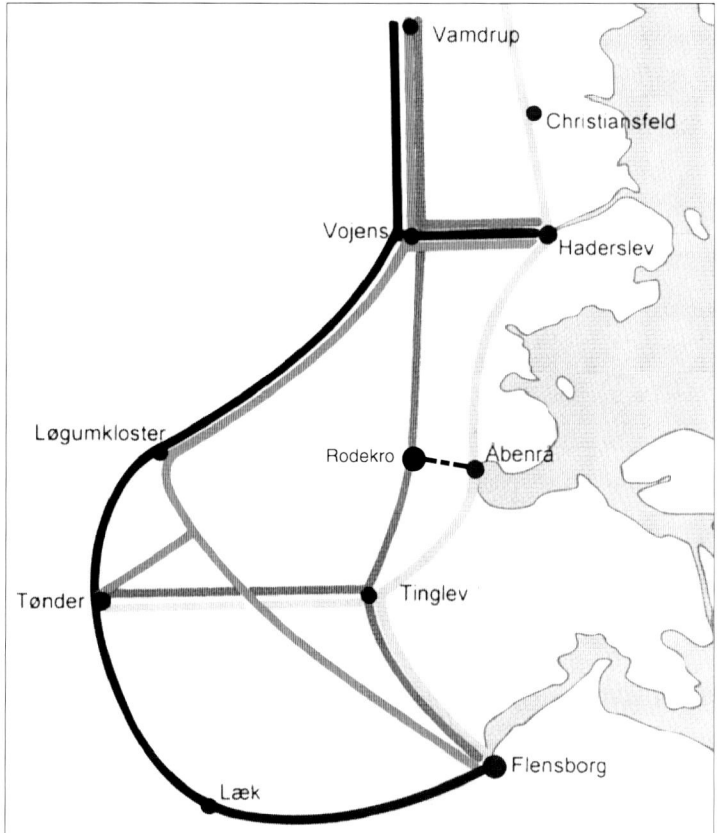

Fig.8:
**Alternative routes proposed by Peto from Flensburg to the Jutland border and considered by the Danish Government.
The line along the east coast included a short branch from Abenrå to Rodekro (on the route up through the middle). The middle route through Rodekro was eventually adopted.**

THE GRAND TRUNK RAILWAY OF CANADA

ONE of the advantages of having become an MP in 1847, was access to those in authority, when railway matters related to the colonies came up for discussion. As part of 'the railway interest' in the House of Commons, Peto was one of the members whose support was sought by those from the dominions with projects that needed approval, and usually financial support, from the Colonial Office. One such emissary was Joseph Howe, a prominent Nova Scotian politican, who came to London in the summer of 1851, to solicit official support for a railway linking the Maritime Provinces with Upper and Lower Canada. If built, this would have been the first trunk line to unite British North America, where there had only been very limited piecemeal railway development. This contrasted with the situation in the United States to the south, whose expanding rail network was beginning to attract traffic from across the frontier and as a result theaten the economic independence of the British territories. Although Howe's entreaties fell on deaf official ears, he did manage to interest Peto, who offered to build the line himself but on condition that the Colonial Office provided half the capital needed and the provincial governments a further £100,000. The money was not forthcoming, however, and the scheme consequently foundered, but Peto made it known that he was open to proposals from other quarters.

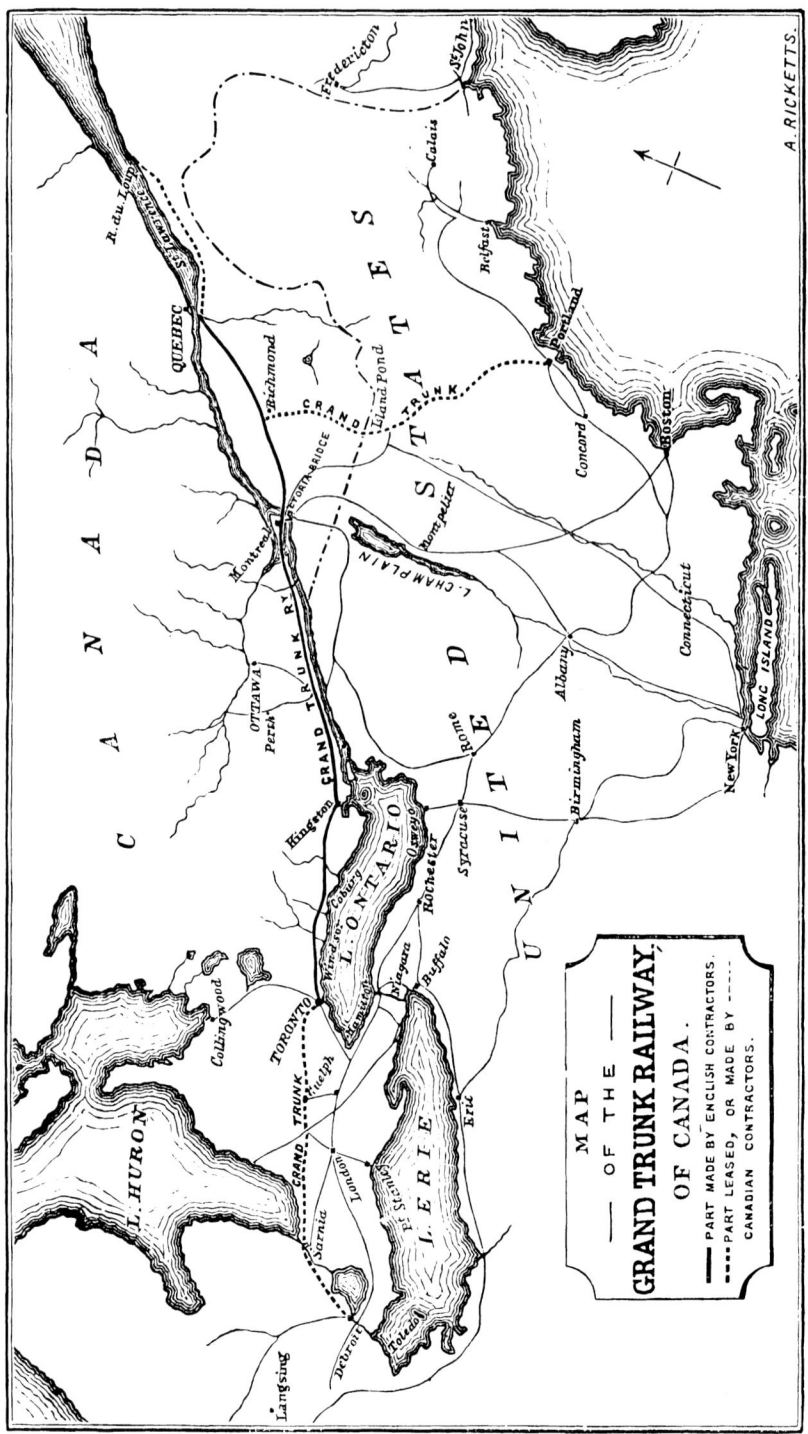

Fig.9: The Grand Trunk Railway of Canada. Peto, and his partners, altogether built over 600 miles in the 1850s. *Helps*

Such a proposition came the next year, when Joseph Hincks, the Finance Minister of the Canadas, came to London on a mission of his own. Like Howe he hoped for financial backing from the Colonial Office, this time for a line across the two provinces, but again it was not forthcoming. Whilst in London he decided to discuss the idea of a privately financed venture with Peto and some other interested parties, including William Jackson, who like Peto, was both an MP and a director of the Chester & Holyhead Railway (C&HR). Jackson sailed back to Canada with Hincks to carry out exploratory discussions with politicians and commercial interests there. These led within a year to the promotion back in London of the Grand Trunk Railway of Canada (GTRC).

In the meantime Peto had remained in contact with Howe, who was now restricting his railway activities to his own Maritime Provinces. It was in connection with these proposals that Peto despatched his agent James Beatty to Nova Scotia and New Brunswick to carry out preliminary surveys in the early 'fifties. This resulted in the contract for a short extension of the existing European & North American Railway (E&NAR) in New Brunswick. Peto began work on this but withdrew in 1855, saying he could no longer provide the financial support required. By that time the GTRC was absorbing all his available resources. His abandonment of the contract appears to have been accepted gracefully by the company. It would be the first and only contract, which he reneged upon voluntarily.

By the time the prospectus for the GTRC appeared in 1853, Hinck's original plan for a line from Montreal to Toronto had been expanded into a through route, more than a thousand miles long, from Portland on the coast of Maine to Sarnia at the southern tip of Lake Huron. Peto, in partnership with Betts, Brassey and Jackson, undertook the construction of more than 600 miles of this in four separate contracts. The longest was the 345 miles from Montreal to Toronto; the others were about 100 miles between Quebec and Richmond; just over 150 miles from Quebec to Trois Pistoles, where it was hoped to join a line to be built from the Maritime Provinces; and the demanding task of constructing a bridge nearly two miles in length across the mighty St Lawrence river, just west of Montreal. In association with the GTRC project, the English contractors also undertook to upgrade the tracks of two existing companies, the St Lawrence & Atlantic Railway (SL&AR), which linked Montreal with Richmond and the Atlantic and St Lawrence (A&SLR) that was constructing an extension from the SL&AR line to Portland, Maine. Both these railways were in a poor state and required much improvement to deal with traffic from the GTRC, as well as needing to be converted to the 5ft 6in. gauge upon which the GTRC was to be built. These appendages were an essential part of the whole project, which was designed to give the most populous parts of British North America access to the markets of Europe throughout the year. Although in the United States, Maine had two important assets; it was ice-free and the passage across the Atlantic was shorter than that from the American ports further down the coast.

All the contracts taken by Peto and his partners were to be charged for on a per-mile basis, £6,500 for the Quebec and Richmond and £8,000 for the others, except the lines to be reconstructed and the bridge over the St Lawrence, which was priced at £1,400,000, with a proviso that this could be increased up to £100,000 'in extraordinary circumstances'. This presumably meant especially severe weather or other unforeseeable physical difficulties during its construction. Despite its fairly modest cost considering its length, the GTRC was required 'to be superior to any American or Canadian railway now constructed and equal to the first class English railways, so far as permanence and substantiality of the work is concerned'. The contractors were also to provide it 'with stations, fixed and moveable plant and rolling stock'.

The contractors must have expected construction costs to be no higher than at home to have accepted these terms, which is surprising in view of the recent experience of three of them in Norway. The terrain across which the GTRC had to be built, was in places as difficult and the winters certainly more severe. The GTRC contracts were taken, however, when only outline surveys had been made of the route. This crossed large tracts of unpopulated and in parts unexplored territory, which added to the construction costs, as there were no proper roads; whilst the decision to keep some miles inland from the St Lawrence river deprived the contractors of ready access to water transport. Robert Stephenson, the GTRC Consultant Engineer, was responsible for this serious handicap as far as the contractors were concerned. As in Norway he delegated the execution of his plans to his Assistant, in this case Alexander Ross, who had worked for him earlier on the C&HR. In some respects the GTRC, although ten times the length, was a replication of the C&HR, transferred to an undeveloped part of the North American continent. Besides having the same engineers, some of the same contractors and the same architect, Francis Thompson, for some of its stations, it terminated at a transit port and needed a major bridge along its route. Before it had been completed, the GTRC would have another very unwelcome aspect in common with the C&HR.

It had been estimated by the promoters that the GTRC would need to raise in all £9½million to cover the cost of the projected line and compensate shareholders in a number of existing companies it planned to absorb, as well as the promoters of some rival lines, who had to be persuaded to abandon their projects. All these parties were apparently overpaid for their co-operation, adding unnecessarily to the capital cost of the GTRC. Nor was the cost of upgrading the A&SLR and the SL&AR, included in the initial costing of the GTRC. With the Canadian provincial government agreeing to contribute £3,000 per mile, or just under a third of the nominal cost of the railway, it appeared at first sight a more attractive investment than most in Britain at the time; consequently the initial flotation on the London market in 1853 was fully subscribed. But Baring Brothers and Glyn Mills, the London bankers, who had the exclusive right to issue the B shares, which made up half the GTRC ordinary stock, with a par value of just over £1,800,000, held back the issue, fearing it might flood the market. Barings informed Peto on the eve of the main flotation that the contractors would be required to accept

all the B stock. This was an enormous liability, as the B stock was more vulnerable to falls in market prices than the A and consequently difficult to dispose of except when economic conditions were favourable. Peto protested but in the end assented, as otherwise the whole scheme would almost certainly have foundered. It was a decision he and his partners would profoundly regret a year later, when the outbreak of the Crimean War unsettled the markets, reducing the value of all the GTRC stock and especially the B shares.

Before the B stock was thrust upon them, the contractors had agreed to take shares in part payment for the works. In the case of the Montreal to Kingston line, which was designated a separate undertaking and a fixed price of £3million substituted at an early stage for the mileage payment, they agreed to take just over £1million in GTRC shares, only £80,000 of which was to be in debentures, plus £1million in Canadian government securities. Although these carried a guarantee of repayment after 25 years with interest at 6% and were consequently more easily placed, the depression inevitably dragged down their market value along with that of all the other GTRC shares. From 1854 until the GTRC was finished, the contractors were in an increasingly unenviable situation, being compelled to accept payment in depreciating stock, whilst at the same time being required to pay out cash for materials and labour. The GTRC compounded the problem by insisting they take a larger proportion of their payments in shares. To make matters even worse, the B stock which the contractors had been obliged to accept at par, had not only depreciated dramatically but when it eventually could be cashed in, was subject to a £12 10s per £100 premium payable to the company.

Although Peto and his colleagues were being compelled to bear an increasing proportion of the depreciation in the company's capital, the GTRC was still in dire financial straits itself, finding it increasingly difficult to raise cash from 'calls' on the stock that had been floated; whilst, as in the cases of the contractors, the government securities they were receiving had to be increasingly discounted – Canadians becoming more disillusioned with the whole project and unwilling to invest. The situation was not helped by scandals over the issue of free shares, said by some of the GTRC's enemies, to have been passed to prominent persons in the colony in return for their support for the scheme. The most controversial case involved Hincks, who was said to have been handed £25,000 of shares but to have paid nothing on them. His political enemies claimed this had been his reward for steering the GTRC Act through the Legislature. Although he denied this, insisting he had not even signed the company's original subscription list, Peto was subsequently obliged to admit that he had entered the shares on the register in Hinck's name and paid the initial deposit himself. Hincks was obliged to resign from the government as a result, whilst Peto's reputation on the other side of the Atlantic was hardly enhanced.

Hincks's fall was not as harmful for Peto as it might have been. A new government headed by Sir Allan MacNab advanced £900,000 to the GTRC and in 1856 passed a Relief Act, which not only permitted the company to issue a further £2million of government guaranteed bonds, but also provided £800,000 in cash towards the cost of completing the bridge over the St Lawrence. Economies were demanded in its construction, however, before the contractors could receive any of the additional payments allowed by the 'in exceptional circumstances' clause in their contract. When Stephenson insisted these could not be made, the extra funding for the bridge was withheld.

The length of the Montreal–Toronto line had necessitated its division into four sections, the contractors putting a separate agent in charge of each. At first they attempted to proceed with these simultaneously but when insufficient progress was made, decided to concentrate their efforts on completing the section from Montreal to Prescott, just to the east of Lake Ontario. The route of the line was also altered in a number of places to facilitate easier construction and reduce costs. Despite the difficulties arising from the terrain and climate, the whole of the line, with the exception of the bridge over the St Lawrence, was completed in 1856. This was an impressive achievement. The Richmond–Quebec line had opened for traffic two years previously but

much still remained to be done elsewhere along the route before the GTRC was operational throughout. Although a lot of the outstanding work was in the hands of Canadian firms, Peto still had to complete the upgrading of the lines east of Richmond, whilst the Trois Pistoles extension had only been taken as far as Rivière du Loup, fifty miles short of its objective, because the projected line from the Maritime Provinces had for the time being been abandoned.

The construction of the giant iron tubular bridge over the St Lawrence just upstream from Montreal occupied 3,000 men, many of them from Britain, for five years. They were obliged to endure excessively-harsh conditions especially in winter and at least fifty died in a cholera outbreak. Designed by Stephenson on the same principle as the Menai Bridge on the C&HR, it had consumed 2¼million cubic feet of masonry and more than 9,000 tons of iron by the time it was finished in 1859. The stone for its twenty-six massive stone pillars, designed to withstand pressure of the ice-flows which sweep down the St Lawrence in spring time, came from a quarry beside the river but still had to be raised into position. This problem was eventually solved with the aid of a mobile crane devised on site by one of the Canadian subcontractors and known as 'Chaffey's steam traveller'. It replaced a machine from England, which had proved to be a failure.

Many of the parts required for the bridge, as well as the locomotives and rolling stock needed to fulfil the terms of the contract, had to be brought across the Atlantic, because there were no facilities to produce them in Canada at the time. Realising the problem at a very early stage, the ever-resourceful Brassey had despatched some of his operatives to the United States to study prefabrication techniques and made extensive use of the system. The Canada Works at Birkenhead was set up by the contractors to manufacture materials for the GTRC. It continued to produce locomotives and rolling stock until the 1870s, supplying a number of the other overseas railways built by Peto and Brassey, as well as some of the railways in Britain in which they were involved (*plate 21a*).

Despite their ultimate triumph in completing Victoria Bridge, which was formally opened and named by the Prince of Wales on his 1860 tour of Canada, the contractors sustained very substantial losses on the GTRC contract. Brassey is recorded as having put the figure as high as £1million, which would have included the depreciation on the shares they handled and the shortfall from the works. Peto must bear most of the responsibility for the former and in particular the fatal decision to accept the B stock, although Jackson appears to have adopted a cavalier attitude on costs and created the impression whilst in Canada that he and his colleagues had limitless resources. They were all inadequately briefed on the construction costs, placing undue reliance upon Stephenson's flawed initial surveys, but should have been aware that labour costs were much higher in North America than at home. There was apparently little attempt until a fairly late stage to mechanise operations on the ground to offset this, despite the fact that steam excavators had been extensively, if not entirely successfully, used for some time in the neighbouring United States.

Peto did not visit Canada when the contractors were obliged to seek additional funding from the provincial government to keep afloat – having made himself too unpopular there to be their spokesman. The less controversial Brassey was entrusted with this task in 1855 and it says a lot for his tact and negotiating skill, that the mission was a success. Although there appear to have been no recriminations between the partners, they all had every reason to regret having become involved in the project and Peto probably most of all. At the time of the opening of the GTRC he wrote:

> They have subscribed £8,000 to spend
> rejoicing in Canada. I don't think if I
> were there, I could have any spirit …
> How keenly I feel the mistake I have
> made, no one can tell.[53]

The GTRC had been Peto's first major contracting failure. To rub salt into the wounds, after its completion, he had to endure a barrage of criticism from those who claimed the contractors had padded out their accounts, by carrying out unnecessary additional work and skimping much of the basic construction, leaving the company to operate a ramshackled line. There was certainly abundant evidence to justify this view of the lines east of Richmond which he had 'upgraded' but these had been in very poor condition when he took them over and, as elsewhere, the cost of the work involved had been underestimated. An independent survey, carried out on behalf of the shareholders by Captain Galton of the Board of Trade, was on the whole complimentary about most aspects of the construction of the GTRC, however, as was a report by S. P. Bidder, brother of George Parker Bidder, although some might have suspected his credentials. For many in Canada the bankruptcy of the GTRC only a few years later, was final testimony to Peto's rôle in the formulation and execution of the project.

THE CRIMEAN RAILWAY

THE outbreak of hostilities between Britain, in alliance with France, Turkey and Sardinia, and Russia in 1854, did have one beneficial result for Peto, providing him with the opportunity to construct the first railway in the World designed specifically to serve the needs of an army in the field. After despatching troops to the Crimea in the autumn of 1854 and establishing a base camp at Balaclava, the allied forces had moved inland to besiege the fortress of Sebastopol but before the end of the year were experiencing serious difficulties transporting men and materials up to the front lines. The only existing road from the coast soon proved inadequate and became a quagmire with the onset of the autumn rains, making it necessary to manhandle most of the munitions needed by the batteries surrounding Sebastopol. When news of this and the other failings of the military command, reached the public at home, there was an immediate outcry and a demand that something should be done at once to rectify the situation. As a MP on the government side of the House, Peto was in an influential position to press his own remedy for the transport problem, upon a by then beleaguered administration. After first obtaining the support of Lord Palmerston, the most influential member of the Cabinet, he presented his plan for a railway to the Secretary of State for War, the Duke of Newcastle.

Peto's proposal was accepted by the Government with the previously agreed proviso that he would build the line at cost and make no profit from the contract, the War Office in return agreeing to pay in cash for all the work. The Crimean Railway was executed jointly with Brassey and Betts, which put the largest and most experienced railway contracting organisation in Britain, at the disposal of the nation.

The first task facing the contractors was to assemble the necessary workforce and materials, then to transport them more than a thousand miles by sea to the scene of action. This operation was masterminded by Betts, who assembled the necessary supplies at the recently completed Victoria Docks in London and commissioned a flotilla of twenty-three vessels to sail to the Crimea. An advertisement in the newspapers for labourers attracted a large response, motivated mainly by patriotic zeal, although the high wages being offered must have been a considerable additional incentive. Even so, it was not possible to recruit enough men of the necessary physical calibre and some of the labourers who had gone over to Canada to work on the GTRC, had to be brought home. About 900 men were recruited and all engaged on a six months contract and provided with free passages. Their rates of pay varied; labourers received five shillings a day, carpenters seven shillings and sixpence and the gangers in charge of the navvies, on an average fifteen shillings. These were significantly higher than what was paid at home even in boom conditions, but this did not concern the contractors as the Government was footing the bill. There was, it must be accepted, an element of danger money involved, although Peto insisted from the start that all his operatives must be treated as non-combatants, informing them he would not be responsible for any injuries they might sustain as a result of the armed conflict

The announcement of the expedition's imminent departure was a morale booster for those at home, Peto no doubt enjoying the favourable comments made about him and his partners in the press. *The Illustrated London News* was fulsome in its admiration of the organisational abilities of the trio of private entrepreneurs, which it contrasted with the failings of the civil servants and the military:

> The public will hail with much satisfaction the departure of the government … from the usual routine of office and red tape traditions. Messrs Peto, Brassey and Betts will execute their business in a business-like manner. They are not the likely men to land their workpeople without tents and tools, to fill the hold of a vessel with medical stores and put tons of shot and shells over them, any more than they would lay the rails of a line and then tip an embankment on top.[54]

The organisation throughout appears to have been exemplary, no doubt helped by the contractors' current experience shipping large quantities of materials across the Atlantic, as well as their contacts with the suppliers of railway materials. The basic needs of the workforce were also provided for. Betts, realising that some form of temporary shelter would be required whilst permanent barracks were being erected, despatched an ample supply of tents and tarpaulins. Peto later claimed he had offered to provide these to the Army as well but met with objections. As a result, the troops were left to sleep in the open, whilst the navvies were protected from the elements.

Peto put James Beatty, recently back from Nova Scotia and long since forgiven for his failings with the tunnel at Southampton,[55] in charge, at a salary of £1,500 p.a. This was comparable with the sums earned by the leading railway engineers in Britain. He was assisted by twenty-five other salaried staff, including four assistant engineers at £500 p.a., a draughtsman at £300 p.a., two surveyors at £250 p.a. and £150 p.a. respectively and a timekeeper (needed because the Government had agreed to pay for the works on a schedule of prices that included labour), at £150 p.a. Also included in the expedition were a surgeon at £500 p.a., four nurses at £50 p.a., which was incidentally the same rate as Florence Nightingale's recruits received after a year of active service, and two missionaries, who at £14 and £8 p.a., were presumably not expected to have many worldly needs!

Beatty proceeded overland to Turkey ahead of the main party and, on arriving at Balaclava, made the necessary arrangements with the military authorities to clear a site beside the harbour for a railway depot. Although the Army assisted him with this task, he refused to allow his navvies to help dig trenches when asked, although they did assist in the sinking of wells and also made lime kilns for the use of the military. By March 1855 nearly five hundred men were at work on the railway, nearly all from England (plate 22a). Some Croatians were employed for a while, but Beatty found them 'worse than useless' for the work involved. With no services available locally, the construction party had to be self-sufficient and for that reason contained a full compliment of skilled operatives and ancillary staff, including seventy-six carpenters, twenty-two smiths and twenty-one masons, as well as eight engine fitters, a well sinker and even a hairdresser. Fourteen engine drivers were also recruited, needed to operate the railway until the Army's own transport corps arrived.

Although the railway was small compared with most of those Peto and Brassey had built elsewhere, amounting, with several short branches, to just under forty miles of double track, and work went on seven days a week, it was not built as quickly as the military would have liked. Nor was it easy to operate, on account of the steep gradients needed to climb 600 feet from the coast to Sebastopol. Stationary engines and ropes had to be used to haul the trains up the 1:15 incline at Kadikoi at the beginning of the ascent from Balaclava, this system being similar to that employed fifteen years before on the London & Blackwall line.[56]

The railway neared completion in August 1855, making it possible for most of the workforce to be sent home, although some, not finding the conditions to their liking, had broken their contracts and departed earlier. There was some justification for this because, as in Canada, insanitary conditions had caused a serious outbreak of cholera, which claimed the lives of sixty men, including one of the missionaries.

Beatty was a casualty himself, although not from cholera. He had stayed on after the rest of the party had left, to assist the Army Transport Corps work the line. Whilst riding down the Kadikoi incline, the brakes on his train failed, causing him to be thrown onto the line. The injuries proved fatal and he died back in England in March 1856. With difficulty Peto managed to obtain a pension for his widow from the War Office. What compensation the contractors provided themselves, if any, is not known, although they did erect a tombstone on his grave in Kensal Green cemetery, testifying to his work in constructing the Crimean Railway.

The obligation in his contract to execute the works at cost did not prevent Peto from using his own plant, or engaging the services of companies in which he had an interest. Before long his critics at home were accusing him of profiting from his patriotic gesture. It was claimed in some sections of the press that he had charged the War Office high prices for inferior materials, defunct wagons from his Chipping Norton Railway being cited as an example. General Sir William Codrington certainly went as far as complaining to the government about these wagons as well as some from the Tilbury line, which had different buffer heights from the rest. The official records certainly show that the War Office paid substantial sums for the hire of vessels belonging to the North of Europe Steam Navigation Company.

When Peto took the Crimean contract, protocol obliged him to resign his seat in Parliament but he speedily received, what probably seemed to him at the time very adequate recompense. In February 1855, as work was in progress on the Crimean Railway and before any rumours had circulated about his indirect profits from it, he received an official communication from the Queen, which stated that:

> Her Majesty is disposed to confer upon you
> the honour of a baronetcy in consideration
> of the liberal and patriotic conduct recently
> evinced of you in the promotion of Her
> Majesty's service in the Crimea.[57]

On the cessation of hostilities in 1856, the now redundant railway was dismantled and the materials offered for sale. Never one to miss an opportunity, Peto duly purchased some of the rails, intending to use them on the railway he hoped to build for the Turkish government from Smyrna to Aydin in Asia Minor. Although this contract did not materialise, he probably made use of the rails elsewhere.

EUROPEAN WORKS AFTER THE CRIMEAN WAR

THE same year as hostilities ended in the Crimea, Peto, again with Betts and Brassey, was engaged by the Austrian Emperor Francis Joseph, to construct just under fifty miles of the Kaiserin–Elisabeth Bahn, which was planned to link Vienna with Linz. The section they were entrusted with ran from Mölk to Linz. The railway was part of a plan to remedy the country's backwardness in railway facilities and, as far as can be ascertained, the contractors were not required to provide any financial assistance. It was the first of several contracts Brassey executed in the Austrian Empire, but the only one he carried out with Peto.

The future German Empire did however provide one small-scale opportunity for Peto. His involvement with the Franfurt-am-Main to Bad Homburg line was, however, entirely financial, the construction being carried out by a French firm, Claussie, in 1859. The lines which Peto built in Schleswig, as pointed out earlier, were subsequently incorporated into the Prussian system.

The Kaiserin–Elisabeth contract did not enable Peto to benefit from the speculation in new railway projects along the lower Danube valley that followed the end of the Crimean War in 1856. His interests in Turkey at the time did, however, put him in contact with J. Trevor Barkley, an English speculator, who was surveying lines for the Ottoman Empire. Through him Peto obtained the contract for a 140-mile line from Ruse, on the lower Danube, to the Black Sea port of Varna. Although the Turks, who still occupied this part of modern Bulgaria at the time, supported the project mainly on strategic grounds, it also had a sound economic purpose, which was to avoid the need for ships to navigate the difficult waters of the Danube delta. From the start the Turkish government agreed to guarantee interest on up to £2million of the company's capital, some of which was almost certainly subscribed by the contractors. Brassey did not join Peto and Betts on this venture, however, but they did manage to persuade their future collaborator in the ill-fated LC&DR scheme, Thomas Crampton, to become a partner in it. Work began in 1862 and the whole of the line was completed by the autumn of 1866. Although the Turks soon decided to restrict its use, the contractors do not appear to have suffered as a result, Crampton subsequently claiming they had made a profit of £250,000 on the undertaking. Nor did Peto's involvement with the Turks, prevent him from obtaining a contract from their enemies, the Russians, who engaged him in 1865 to build 162 miles of track from Dunaberg to Witepsk. Unfortunately the crisis at home prevented him and his partners from completing this.

Despite the protectionist policies pursued in France after 1848, Peto still hoped to be allowed to continue with the railway construction he had begun with Brassey and made no less than fourteen visits to Paris in the twelve month period from December 1859. The only positive outcome was the commission to construct thirty miles of track in Algeria. The line was the first in the colony and ran from the capital Algiers, inland to Blida. A start had aleady been made using military labour but this soon proved unsuitable, at which stage Peto was called in. He went out to Algiers in 1860 to attend the official inauguration of the scheme, which included the laying of the foundation stone of the station at Algiers and a ceremonial dinner. Unfortunately due to an oversight, he was not invited to the dinner, which must have hurt his pride somewhat. When Napoleon III got to hear of the *faux pas*, however, he did his best to assuage Peto's feelings, by granting him a private audience the next day.

Although the Algerian railway was completed to the satisfaction of the French authorities, growing political tension between France and Britain in the early 'sixties made it difficult for Peto to obtain further work in Algeria or France itself. He did his best to counter the anti-French hysteria which swept Britain at this time, advocating a conciliatory policy on several occasions in the House of Commons. Here he was not motivated entirely by contractual considerations, being a genuine admirer of France. He expressed this in his strong support for the commercial treaty negotiated by his parliamentary colleague, Richard Cobden in 1860 and his opposition to Lord Palmerston's decision to erect defensive forts in the Solent to repel an anticipated French invasion.

Peto's Francophile sentiments did not extend to the activities of the French railway contractors, who apparently deprived him of a substantial contract he had hoped to obtain in Portugal in the late 'fifties. In connection with this he made a number of visits to Lisbon. A draft contract was actually agreed between his agent and the Portuguese government in 1857, which would have allowed him to complete the Northern Railway, upon which work had already begun, and extend this southwards to the capital. According to the notes which survive in the Portuguese official archives, the arrangement would have been for:

> The Government to pay to Sir Morton Peto the whole subvention £1,807,000 voted for the line, in monthly instalments as may be arranged … Sir Morton Peto to pay to the Government six per cent per annum interest of a sum not exceeding £120,000 until the whole line is opened and the £120,000 to be repaid … as hereafter provided.
> A company to be formed as soon as possible but if not done … Sir Morton Peto to take possession of the whole line … and work the same paying the Government twenty-five per cent of the gross receipts.

Despite lengthy discussions and visits to some of the works that were in hand, during the course of one of which he was loudly cheered by the navvies, Peto's plans came to nothing. He later attributed his failure to his refusal, unlike his French competitors, to bribe the Portuguese officials. In the end he had to be satisfied with a ceremonial send-off from Lisbon harbour, when his vessel was escorted by three royal barges.

THE ATLANTIC & GREAT WESTERN RAILROAD

ALTHOUGH Peto had done no further work in Canada after completing the GTRC in 1860, he was involved with another, almost equally ambitious scheme, only two years later, in the neighbouring United States. The Atlantic & Great Western Railroad (A&GWR) displayed some of the characteristics, as well as weaknesses of the GTRC. It was also built on an unconventional gauge, in this case 6 feet, when the 4ft 8½in. standard British gauge was already predominant in the United States. In conjunction with lines on either side of it, the A&GWR formed part of a projected direct rail route between New York and St Louis, although it only itself ran from Salamanca, in Upper New York State to Dayton, Ohio. It also contained a branch from Meadville to Oil City in Pennsylvania, to serve the embryonic oil Industry, and another to Cleveland, Ohio, its total length being 605 miles. Unlike the GTRC, the A&GWR traversed territory already served by railroads, which offered competition for through traffic without being easily able to act as feeders, having mostly been built on the 4ft 8½in. gauge.

When first put forward in the 1850s, the A&GWR project had failed to attract sufficient backers and was shelved. Its subsequent revival was mainly due to the efforts of James McHenry, a former Liverpool cotton merchant, who had turned to railway speculation. He managed to attract the interest of a number of British capitalists and a London Board of Management was duly set up. Despite his recent experience with the GTRC, Peto had great confidence in the future prospects of the United States and invested heavily in the company, soon becoming its London Chairman. His firm was never, however, involved in the construction of the line, McHenry acting as contractor, although he did carry out the work in collaboration with Smith & Knight, a firm Peto had had dealings with previously on his side of the Atlantic.[58] The engineer was T. W. Kennard, an Englishman, also familiar to Peto, who had gone to work in America some while before.

Although the company managed to raise sufficient funds to proceed with the construction of the line in 1862, it soon found itself unable to raise all the capital needed. It was also obliged to suspend work for a while, due to the shortage of labour, caused mainly by the American Civil War, which had broken out the previous year, many of the potential navvies having been drafted into the Union Army. Peto was a strong supporter of the Union, not only on account of his investment in the A&GWR, but because, like most of his fellow nonconformists in Britain, he was strongly opposed to slavery.

Fortunately for the A&GWR, the districts through which the line passed were well to the north of the combat zone, whilst the labour shortage was partly relieved by recruiting navvies formerly employed on the GTRC. The line was completed in 1864, but traffic failed to come up to expectations, making it a liability for the shareholders. The London Committee soon decided that the best remedy was to either sell or lease the concern to another American railroad.

With this object in mind, Peto crossed the Atlantic in the autumn of 1865. After a spell in New York and Washington, where he met President Johnson and General Mead, the victor of Gettysberg, he travelled the whole length of the A&GWR inspecting the line and its facilities. These included the repair shops at Corry and locomotive works at Meadville in Pennsylvania. Also sited at the latter was a 100-bedroom hotel, built in the English style, which, it was reported at the time, had the best dining room in the whole of the United States. There is no evidence that Peto invested in the hotel but he certainly sampled its fare before proceeding on his journey. He failed, however, in his main objective of disposing of the struggling A&GWR, which would prove an additional liability for him in the crisis of the following year. During the bankruptcy proceedings he claimed that McHenry owed him over £100,000. Whether this was correct and if so, how much of this he got back, is not known. The company for its part, was finally driven into receivership in 1874.

AUSTRALIA AND ARGENTINA

BY THE end of the 'fifties Peto's joint overseas contracting operations with Brassey, had taken on a worldwide character. In 1859 they began their first Australian railway, with fifty miles of track in New South Wales. But despite supplying the ironwork for the Stony Creek bridge on the Melbourne to Williamstown Railway in neighbouring Victoria, they failed to secure the contract for the proposed line from Melbourne to the Murray river. They did, however, build some further lines in New South Wales, as well as eighty miles of track in Queensland, during the mid-'sixties. All the Australian work appears to have been put into the hands of Brassey's agent, Wilcox, which suggests that Peto's involvement was entirely financial.

Although Peto and Brassey both undertook contracts in Argentina, they carried these out independently. In 1864 Brassey obtained a contract from the Central Argentine company for nearly 250 miles of line. Peto had been engaged the previous year to build the first, seventy-two mile long, section of the Great Southern Railway (GSR), which ran from Buenos Aires to Chascomus and again took the form of a government concession. Although the GSR directors were subsequently accused of bribing government officials, there is no evidence that any of Peto's staff were involved.

By any standards Peto's overseas contracting achievement is impressive. In all he constructed over 1,700 miles of railway abroad, more than double his total track mileage at home, the works extending over five continents and occupying him continuously for fifteen years. All except about 400 miles were joint operations with Brassey, however, and besides leading the way, his partner was clearly an indispensible ally on a number of occasions, and in the case of the GTRC, their largest joint enterprise, possibly his financial saviour. Whilst it is not possible to estimate either the total extent of Peto's financial involvement in the foreign ventures, or the return from them, the losses on the GTRC and the A&GWR undoubtedly substantially reduced the overall return.

NINE

Partners, Agents and Workforce

PARTNERS

EXCEPT for the short periods when he worked on his own, Peto conducted his business as a partnership. This was the way most enterprises in Britain were organised before the growth of the joint stock company quite late in the nineteenth century. The railway companies, like their canal predecessors, were an exception, being pioneers of the new form of enterprise, which was particularly attractive to smaller investors, as it limited liability to the face value of the shares purchased. Railway contractors on the other hand had no such security, being liable in law for all their commitments, as was painfully evident in Peto's case with the LC&DR contract. They not only needed capital to start work but were obliged to risk this before they could be certain of any profit. Peto embarked upon the business with two advantages over most of his competitors: he had very substantial resources of capital and his firm had another line of business to offset the risks. The men who entered into partnership with Peto shared the risks and the profits, although in what proportion it is impossible to say, in the absence of any surviving records. Like many others, his closest and longest lasting partnerships were with members of his own family. This was particularly appropriate in his case, as the source of his initial wealth had been his uncle's legacy.

Peto constructed his first 250 miles of railway in partnership with Grissell and appears to have been given a largely free hand in the execution of the work, even before the completion of the Great Western contracts. This indicates not only his partner's willingness to entrust him with large and risky works, but also the absence of joint overall management of the firm, which could have grown too large for that by this time. They may have consequently decided to divide up the contracts and take responsibility for each individually, together with the profits and any losses, evidence for this being Grissell's refusal to become involved in Peto's dispute with the Great Western. It was probably only when Peto threatened the solvency of the firm by accepting large amounts of railway shares in payment for the works, that Grissell got cold feet. Although this tension began in the early 'forties, the partnership was not finally dissolved until 1846, which suggests he could not have worried too much at first. It is possibly more significant that they eventually parted ways at the height of the Railway Mania, when Peto had taken on more commitments than ever before. Grissell may have foreseen the eventual outcome of the wild speculation at the time and much preferred to concentrate his resources upon the lucrative and safer Houses of Parliament works. These would in due course provide enough return for him to retire to the country. According to Peto, the dissolution of the partnership was amicable on both sides. Evidence for this is Grissell's re-payment of a sum found still to be outstanding to Peto long after the final settlement of their accounts.

Grissell had been eight years older than Peto. His second partner, Edward Ladd Betts, was six years younger, although he had begun railway contracting slightly earlier. Together they constructed over 500 miles of track in Britain, almost exactly double the mileage Peto completed with Grissell, although they only worked together for six years longer. In addition Betts partnered Peto in the construction of nearly 1,700 miles abroad. The exact nature of their working arrangement, like those with the other partners, remains obscure, although Peto's son Henry, in his posthumous biography, says that the supervision of work on the ground was normally in the hands of Betts. This left Peto free to concentrate on financial matters, which was his main preoccupation throughout the period they worked together. Betts invested less than Peto in the railway companies which they jointly promoted and was not as much involved in their management, holding fewer directorships. The two he did hold seem to have been as Peto's 'front man'. The first was as early as 1845, before they had entered into any formal partnership arrangement, when he joined the Board of the Eastern Counties Railway (ECR). If Peto was owed money by the ECR at that time, he would have wanted to make his influence felt but was prevented seeking office by his ongoing contracts. The other occasion came after the completion of the Hereford, Ross & Gloucester Railway (HR&GR), when the contractors' investments justified representation on the Board, but Peto was probably too busy with the Chester and Holyhead and other commitments to give the time.

Thomas Brassey was an even more important figure on the railway contracting scene than Peto & Betts. By the time they joined forces in the early 'fifties, Brassey had acquired an enviable reputation for efficiency and reliability going back to the time he had taken over the construction of the trouble fraught London and Southampton line a decade earlier. The majority of their joint operations, 1,300 miles in all, were carried out abroad, although they did build 120 miles together in England, with significant financial involvement on both their parts.

The Peto/Brassey co-operations appear to have been *ad hoc* arrangements with no formal long-term commitments on either side, and probably as much financially as operationally motivated. There is some evidence, although by no means conclusive, that the joint contracts were individually executed rather than shared on the ground, with the contractor whose agent directed operations assuming responsibility for the construction of the line. On this basis, Peto carried out more of the joint works in Britain, and Brassey by far the greater number abroad.

The absence of a formal partnership made it easy for Brassey to disengage himself from Peto and not become involved specifically with the LC&DR contract, although the two continued to co-operate abroad in the early 'sixties. As a result Brassey managed to survive the 1866 crisis and take over Peto's interests abroad. It is conjecture whether Peto would have survived, had he carried out the LC&DR works jointly with Brassey, as happened with the GTRC, or if both would have gone down, assuming that Peto had remained in charge of the finances. Whilst there was never any chance that the LC&DR would be reprieved by government intervention, Brassey might have been able to curb some of the excessive risks taken by Peto.

Peto's other partners are far less significant. Little is known about William Easted, who joined him at the start of the Birmingham and Oxford Junction operations.[59] William Jackson, the fourth partner in the ill-fated GTRC contract, was a financier more than a contractor, who managed to avoid commiting himself too deeply in the venture, enabling him to extract himself comparatively unscathed to die a wealthy man. The third contractor in the LC&DR tragedy, Thomas Crampton, was a late recruit to the contracting organisation and, like Betts, probably did not involve himself greatly in the financial arrangements. Evidence for this is Peto's use of the proceeds of their joint Varna operation for his own purposes, Crampton telling the Bankruptcy Court that this had been done without his knowledge, which he appears, on the surface at least, to have accepted. Peto seems to have been given a virtual *carte blanche* by his partners on matters financial, or to have managed to conceal the real state of affairs from them.

ORGANISATION

ALTHOUGH Grissell & Peto's operations, including the early railway undertakings, were managed from the firm's Lambeth headquarters, the need to direct personally the railway works made it necessary for Peto to adopt a peripatetic lifestyle during the last five years of the partnership. He took short-term leases of properties in Hertfordshire and Norwich to oversee the execution of the ECR and NR contracts. Although the expansion of the rail network and increasing delegation of the supervisory work, made frequent changes of address unnecessary by the end of the Mania, his lifestyle became, if anything, more rather than less mobile. His eldest son Henry, in the course of sorting through the surviving family papers prior to writing the biography of his father, found letters that Peto had written on successive days in 1847, from Southampton, London and Norwich. The foreign contracts of course necessitated longer journeys. Besides the crossings of the North Sea, there were voyages to Lisbon, North Africa and finally to the United States. But wherever he travelled in later years, it was as a businessman, rather than the supervisor of operations. Another indication of his changed role after the break with Grissell was his decision to move his business headquarters to the north bank of the Thames in Great George Street, Westminster, where many of the leading railway engineers had their offices. It was also conveniently close to the House of Commons.

The key to Peto's success as a railway contractor in the early days, was an efficient organisation on the ground, with a well-defined chain of command, from himself at the top to the humble navvies who were carrying out the work. The size of his undertakings necessitated the employment of a very large labour force, which in the age before any significant mechanisation, was the most important element in the construction process and accounted for a very large proportion of the contractor's costs. At the height of the Mania Peto was employing directly or indirectly more than 9,000 men so it is particularly fortunate that he has left a fairly detailed account of the way he managed his workforce at that time.

In May 1846 Peto was summoned, along with a number of other contractors, to give evidence before a Parliamentary Select Committee that was examining the conditions of those employed on the numerous railways then being constructed throughout the country. The Committee had been set up in response to public concern at the harsh and dangerous nature of the work and the lawlessness of many of those engaged in it. Realising the spotlight was upon him as a major employer, Peto set out from the start to give the best impression possible of his own organisation.

He was building four railways at the time – the Southampton & Dorchester, the Wymondham branch of the Norfolk, the Lowestoft and the Ely & Peterborough, but had to admit that he was making extensive use of subcontractors, almost all his own labour force of 3,700 men being tied up with the last two contracts. He told the Committee:

> It would be almost impractical, nor would it be desirable, to have the onus of carrying these large works out entirely myself … without availing myself of the assistance and talents of those men who have made some money in my service or others who may have capital.[60]

He might have added that they shared the risks and avoided the need for him to expand his own workforce at a time when labour was in short supply and expensive; also they could be easily discarded, when the construction boom came to an end.

The subcontracts varied from as much as £30,000 for several miles of track to a few hundred pounds for a length of embankment which might be let out to a former ganger who had accumulated a little capital. After agreeing the price for their work and undertaking to be supervised by his agent, the subcontractors paid Peto a ten per cent deposit, which was a similar arrangement to his own with the railway company and its engineer. Like him, the subcontractors were also paid monthly for the work they did, although those with limited capital might receive advances at more frequent intervals. The smaller subcontractors were usually limited to excavation and earth moving. Even on a largely subcontracted line, Peto said he usually reserved the technically more demanding work, like the erection of swing bridges, for his own staff. He was nevertheless prepared to sub-let the tunnel at Southampton on the Dorchester line.

The subcontractors, like Peto himself on the works he carried out with direct labour, sub-let their undertakings to gangers, who agreed a price 'by the lump' for the work they carried out. This spread the risk further down the chain of command, with neither Peto, nor any of his subcontractors playing any part in these negotiations. There was one proviso in all cases, however; every man employed had to be paid his full wages weekly in coin of the realm, a clause to this effect being written into all the subcontracts. The subcontractors and gangers were also forbidden to sell goods or make payments in kind to anyone employed on the works. The 'truck system' was thereby effectively banned from all Peto's works, and he told the Committee he was one of only a very few railway contractors insisting upon this.

By 1846 Peto was delegating the day-to-day supervision of the works to his appointed agents. This select body of men stood at the top of his staffing hierarchy and enjoyed salaries commensurate with their status in the firm; although the £1,500 per annum that Beatty received when put in charge of the Crimean contract, was probably well above the norm, as the Government was footing the bill. Little is known about the background or training of any of the agents, although one, Mr Rowan, was a qualified civil engineer. He stayed with Peto for more than twenty years, starting on the East Lincolnshire line in 1848, then crossing over to Canada to take charge of a major portion of the GTRC, and ending up with an important post in Denmark. Beatty, described by Peto as 'the fair haired young man from the North of Ireland', who had begun working for him on the Yarmouth–Norwich line, remained with the firm until his untimely death. Peto must have had complete confidence in his abilities when he put him in charge of the sensitive Crimean

contract and the same applied earlier with the subcontracted Dorchester line. Another trusted lieutenant, Mr McKeone, who had started on the ECR and also worked on the East Lincolnshire, had the far from easy task of negotiating the terms of the abortive Portuguese contracts with the local officials. Peto summed up the regard in which he held all his agents, during the course of his speech at Lowestoft in 1860:

> In all matters in connection with myself and firm … we are but one, activated by one common feeling, having one common end in view and one common character to maintain.[61]

On the lines which were built with direct labour, the agents were responsible for engaging the gangers and agreed the sums they were to be paid for their parts of the work. During the course of his evidence to the Select Committee, Peto claimed that great care was taken over the selection of the gangers and their activities were kept under firm control. To this end a timekeeper was employed every few miles, one of whose responsibilities was to see that the workmen received their dues every week and also that the gangers obeyed the anti-truck regulation. On 'schedule of prices' contracts, his returns were also the basis upon which the railway company was charged for labour.

WORKFORCE

PETO was undoubtedly sincere in his opposition to the 'truck system', describing to the Select Committee the advantages of the method he followed :

> During the last sixteen years I have always paid the men in money and have found the good effects of it in the moral character of the men, in their steady attention to the work and my own ability in consequence of carrying out works more creditably to myself and satisfactorily to my employers, than I could have done by any other system.

He had to admit, however, that 'truck' was endemic to his industry, many contractors looking upon it as a means of supplementing their profit from the works. There were practical problems to be overcome before it could be suppressed, in particular the difficulty in provisioning a large workforce in remote areas. But he pointed out how he had managed to do this when building the Ely & Peterborough line, which crossed some of the least accessible parts of the Fens, telling the Committee:

> My agent … would go to the nearest market town and say at such a place or such a place, I shall be paying away £500 or £600 a week to the men and it will be in your interest to take care that the men are well supplied; he would go to Ely, Peterborough, Whittlesea and instigate the different parties to come with a good and abundant supply on Saturday afternoon. At one place I saw several butcher's carts loaded with meat and the butchers' men crying out: Who wants a fine leg of mutton? And there is great competition to supply the men. At Ely you would see thirty or forty baker's carts, all piled high with bread going into the Fens on Saturday to supply my men.

Peto was of course employing direct labour here. When it came to the subcontracted works, the eradication of 'truck' was more difficult, although a ganger employing a large number of labourers on the tunnel at Southampton, was certainly hauled before the local magistrates, when he attempted to initiate the practice.

Peto outlined to the Committee some other measures he had taken to improve the condition of his workforce. When there were not enough lodgings available in a district, the firm erected temporary wooden barracks to house the men. As the accommodation was let to them at a nominal rent, he could hardly be accused of operating the truck system he so much detested. A woman was also engaged to cook and make the beds for groups of single men, although he was at pains to point out that she was usually the wife of 'a steady man' and that other women were strictly excluded from the premises! There was also a sick club, to which all his employees were obliged to belong. This was nominally run by them, although the firm held the funds on their behalf and the accounts were kept in the agent's office. Weekly payments were made to those unable to work, varying from twelve shillings a week for a mechanic to eight shillings for a labourer, approximately half their normal wages. The scheme was in theory self-financing, although Peto admitted that on occasions he supplemented the funds but concealed this from the men, which he could easily do, as his agent kept the books. In addition he always met funeral expenses if one of his men was killed and also paid the widow a lump sum of £10.

Although there is no evidence that those employed on Peto's early railway contracts resented the compulsory deductions from their wages, there was certainly some opposition in later years. At one of his election meetings at Finsbury in 1859, some of his workmen were reported to have demonstrated their disapproval of what they considered his authoritarian methods. It is not clear whether they reflected the views of a significant number of Peto's employees, or if their protests were politically orchestrated. The navvies had good reason all the same to be suspicious of the financial dealings of contractors; and especially the gangers, who had the all too frequent habit of absconding with their wages. When this happened Peto would, he told the Committee, make up the deficiency after a thorough investigation had been carried out. The natural suspicion of the navvies also accounted for the failure of an attempt in the early days to set up a savings bank; Peto being obliged to recognise that the men were naturally secretive about money, saying they had 'not as yet learned the value of interest'. He cited the case of one of his navvies who had been killed, and was subsequently found to have £50 hidden under his bed.

In reply to a question from a member of the Select Committee, Peto declared that there were no trade unions amongst his railway operatives, unlike their counterparts on some of his London building sites. This was a veiled reference to the major dispute he and Grissell had been involved in five years earlier and which had seriously disrupted work on the new Houses of Parliament, Nelson's Column and the Woolwich Dockyard extension.

Plate 17a:
Lowestoft station in 1991 prior to rebuilding. *J. D. Bennett*

Plate 17b:
The main concourse of the station. *J. D. Bennett*

Plate 17c:
The Royal Hotel, Lowestoft, now demolished. *By kind permission of Lowestoft Public Library*

Plate 18a: **The construction of the LC&DR City Lines.** (*Illustrated London News* 23/4/1864)

Plate 18b: **Thomas Crampton who joined Peto as a partner in the 1860s.** *Science Museum*

Plate 18c: **Work in progress on the Blackfriars rail bridge.** *Institution of Civil Engineers collection*

Plate 19a:
**Crystal Palace High Level
terminus of the CP&SLR,
now demolished.**
(*Illustrated London News* 30/9/1865)

Plate 19b:
Peto in middle life.
*Portrait by Kelsall Peto in the
possession of the Peto family.*

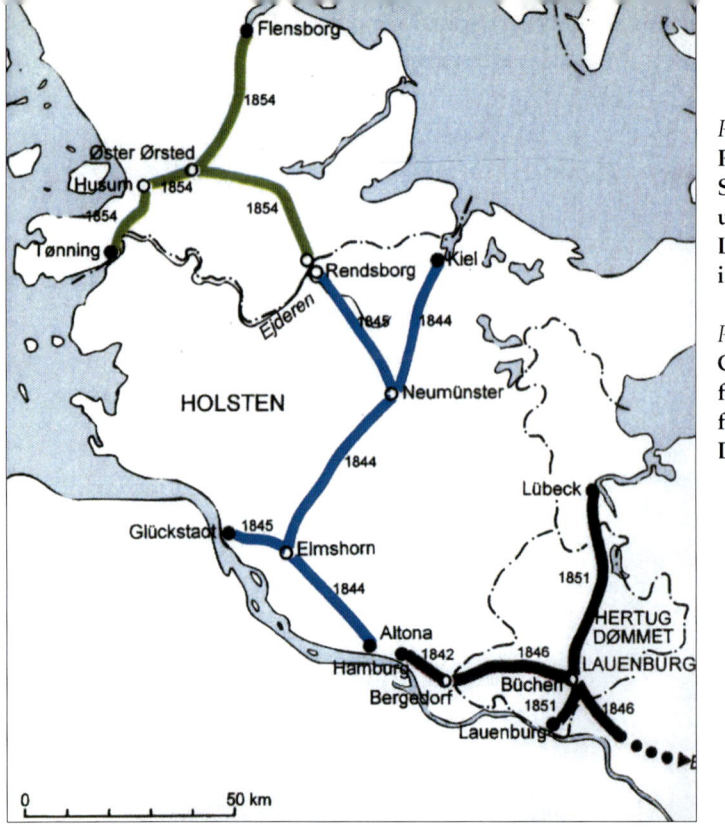

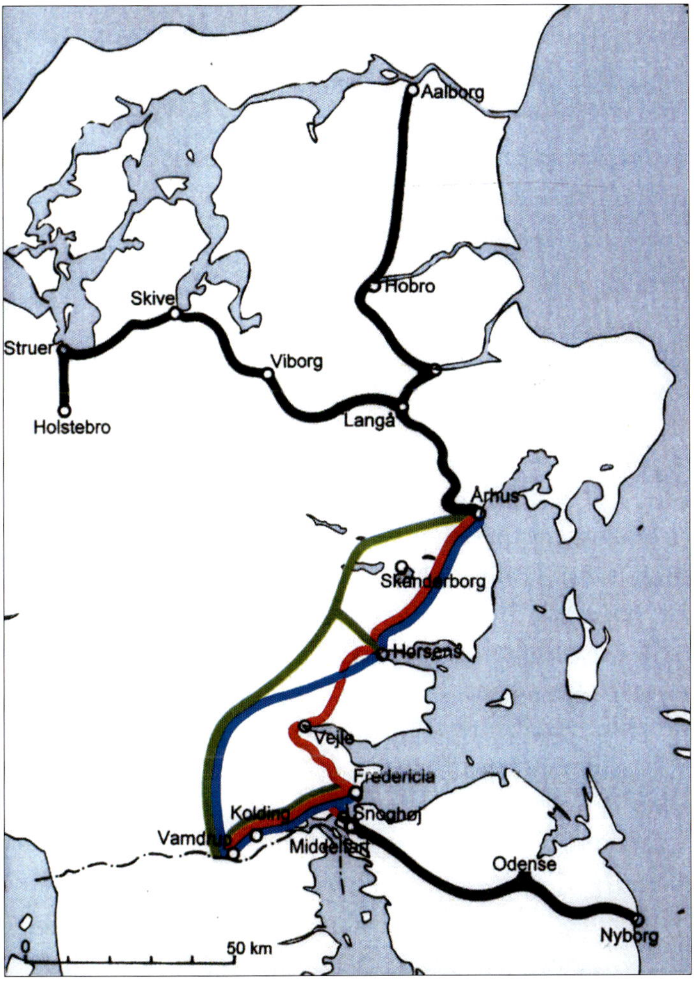

Plate 21a:
Locomotive built at the Canada Works for the Jutland lines at Randers in 1866.

Plate 21b:
Proposals made by Peto in 1860 for railways in Jutland and Fyn, including alternative routes along the east coast of Jutland, which are shown in red, green and blue.

Source for plates 20 & 21 except 20c:
P. Threstrup: *Dampen binder Danmarksammen,* Vol.1., Odense, 1997

Plate 22a: **Commencement of railway works at Balaclava.**
(*Illustrated London News* 10/3/1855)

Plate 22b: **Peto** (*centre*) **photographed with his wife and family, probably at Pinner some time in the 1870s. Henry is the figure with the walking stick at the extreme right; Harold is sitting on the ground immediately below him; the youngest son, Basil, is at his mother's knee.**
Peto family

Plate 23a: **Peto in old age.**
(H. Peto: *Sir Morton Peto*)

Plate 23b: **Bloomsbury Baptist Chapel in 1848.** (F. Bowers: *A bold experiment*)

Plate 23c: **Peto's second town house in Kensington Palace Gardens.**
(J. Summerson: *London building world of the 1860s*)

Plate 24a:
Peto's last residence at Pembury near Tunbridge Wells. He died there in 1889. *Author*

Plate 24b:
Tombs of Peto and his second wife in Pembury Churchyard. *Author*

Plate 24c:
'The Free Church that became a Freehouse'. A 21st-century memorial to Peto: 'The Samuel Peto' at Folkestone was a Baptist chapel before its conversion into a gastro pub. J.D. Wetherspoon chose the name as Peto provided funds to build the chapel. It is an apt choice as the building is very close to the viaduct which Peto had built earlier. *Author*

At that earlier time the partners had been at the centre of a much publicised and vitriolic dispute, memories of which must still have lingered in 1846. This had centred upon the activities of their principal foreman, George Allen, a one-time trade unionist, who by the early 'forties had changed allegiances and become very much the bosses' man. Clearly an able man with original ideas, Allen had designed a new system of mobile scaffolding, which considerably speeded up work. When he introduced it on the Houses of Parliament site, however, he came into conflict with the workmen, and especially the masons, who strongly resented being put under pressure to work faster. Matters were not made any better by Allen's abrasive personality, which earned him the nickname 'The Black Prince', one of his adversaries declaring him to be a man 'who damns, blasts and curses at every turn'. The masons who instigated the strike were the employer's most formidable opponents because, besides belonging to a well-organised craft union they could not be as easily replaced as the less-skilled members of the workforce. When the masons employed at the Houses of Parliament were joined by their colleagues at the other sites, as well as many of the other operatives, Grissell & Peto had a potentially crippling dispute on their hands. Their response was draconian rather than conciliatory, sacking the strikers and replacing them as far as possible with 'blackleg' labour. This did not prove as easy as they had hoped, however, and work was severely disrupted for several months, the solidarity of the strikers being reinforced by the support they received in the Radical press. In the end however, capital prevailed over labour, the strikers failing in their main aim – the removal of Allen – but not before Grissell & Peto had acquired an unenviable reputation in some quarters as a ruthless employer.

The firm had little railway work in hand during the 1841/2 strike, having just finished the Blackwall line and not yet started on the Yarmouth & Norwich. This may have strengthened their will to resist the strikers on building sites. The isolation of many of the railway works and the overwhelming preponderance of the unskilled amongst the workforce, along with the comparatively high pay, probably discouraged trade union activity. Nevertheless the pressure to complete a contract on time, especially when there was a bonus at stake, made it in Peto's own interest to maintain a contented workforce, and this was clearly his policy. The only significant strikes on his railway works were abroad, the most serious on the Victoria Bridge at Montreal. Here the workers were in a stronger position than in Britain, labour being in short supply, with the skilled masons and welders especially difficult to replace. In Denmark on the other hand the labour guilds had legal rights on pay and conditions, which the employer was obliged to respect, although the system excited the wrath of the normally mild Brassey, who like Peto, saw these regulations as unacceptable impediments to the pursuit of *laissez-faire* capitalism.

Peto also claimed to be concerned for the spiritual as well as the material welfare of his men. He told the Select Committee he gave a Bible to every man who requested one, although he had to prove he could read and was required to present it for inspection if requested to do so. Although there was no attempt to set up schools to teach the navvies to read or instruct their children, they were encouraged to attend evening classes, and at Lowestoft these were certainly laid on specifically for them. He also encouraged them to attend religious services and had gone to the trouble of erecting temporary chapels every nine miles on the Norwich–Brandon line. Scripture readers were engaged to lead the Sunday services, because in Peto's view laymen were better able to 'get down to the level of the men' than 'a man who has received a university education', citing his experience on the South Eastern line, where he had engaged an Anglican priest under licence from the Archbishop of Canterbury, who had proved ineffective. Peto by this time had joined the Baptists and his stated preference for Bible readings and 'a plain sermon', rather than the recitation of formal liturgies, complied with his own religious stance.

Whilst he lived in Norwich, Peto had attended St Mary's Baptist Chapel and become friendly with the Minister, the Reverend William Brock. Recognising Brock's abilities both as a preacher and an organiser, he decided to put him in charge of a mission he set up to minister to the workers on his railway sites. Brock selected the scripture readers, known not inappropriately as missionaries, posted them to particular locations and supervised their work. Besides conducting services they were expected to visit the men at home and to minister to the sick. Whilst accepting that not all those appointed had succeeded, Peto assured the Select Committee that on the whole the missionaries had exercised 'a large measure of moral influence' and also helped him to control his workforce. Although not unique by the eighteen-forties, this was a serious attempt on Peto's part to bring religion to his workers. It was a policy he persisted with for some time, being careful to include two missionaries in the party he sent out to the Crimea. One, Mr Gyngell, was amongst those who died there from cholera. Brock eventually gave up his post to become again, at Peto's instigation, the first minister of Bloomsbury Baptist Chapel in London. Although his congregation in fashionable Bloomsbury was very different from the itinerant railway navvies, Brock did instigate some pioneering social work in the slums of neighbouring St Giles during the time he was there.

Peto gave the 1846 Committee some details of the wages those employed on his railway works were receiving at the time. These varied between an average of twenty shillings per week for labourers, to as much as thirty-five shillings for skilled workers, such as masons and smiths. This was for a ten-hour day, although on Saturdays work usually stopped at midday. If the situation demanded, as for instance when Brunel set a deadline for the completion of the Great Western line into London, operations could proceed around the clock, with huge torches illuminating the works at night. Extended working was also quite usual in the summer months, to take advantage of the longer daylight hours and more favourable weather. Peto did not say what the men received for such enforced overtime, but overall the pay was considerably better than for comparable work elsewhere.

Rates abroad could be considerably higher than at home, as the records for the Balaclava Railway show, although these may have been exceptionally high. In Canada, it was necessary to pay well to attract sufficient labour and at one stage those crossing from England had to be given a free passage.

The wages reflected the demanding nature of much of the work, a typical navvy in the age before the advent of large-scale mechanisation, being expected to move ten tons of earth in a day. Peto, like Brassey, appears to have had little use for steam excavators, perhaps because one used on the Colchester line, although not by him, had proved unsatisfactory. The heavy nature of their task called for men of stronger physique than most of the local population. On the Southampton & Dorchester line for instance, their muscle power was said to earn the navvies twice the pay of the average farm labourer, which apparently gave rise to fears of unrest in the rural community.

Although railway construction was dangerous for those engaged upon it, none of Peto's lines, in Britain at least, appears to have had an exceptionally high accident rate. This may owe more to the nature of the terrain crossed than any exceptional safety precautions taken by his staff. In the days before employers' liability was made a legal obligation, contractors were not required to compensate any of their employees for injuries sustained at work, and Peto may have been exceptional in making payments to widows and indirectly supporting sick clubs. He strongly opposed any form of compulsory compensation, however, telling the Select Committee that could be 'a source of great litigation' and he would have to pass the additional costs on to the companies whose lines he was building. In his view: 'no man can be more careful than the contractors are, or more anxious to prevent accidents', but this was small comfort for those who were injured.

As pressure for legislation on industrial injuries mounted in the succeeding years, Peto became increasingly outspoken in his opposition to what, he considered, was unjustified interference in employers' rights. When the first Bill on the subject came before the House of

Commons in 1862, he declared that if it was passed 'masters would be plunged into a sea of litigation, out of which he did not see how they were to emerge'. Whereas in 1846 he had only been mildly critical of some of his workmen, saying they 'sometimes run risks of their own choosing' such as 'working lifts with too high a face', a decade and a half later he accused them of acting on occasions in a totally irresponsible manner. Quoting the case of one of his labourers who had been found sitting on a barrel of gunpowder smoking his pipe, he commented:

> Had that man been, as he [Peto] was, the father of twelve children, and the keg of powder had blown him in the air, he, the employer, would have been very much dissatisfied had he been called upon to support the unfortunate offspring.[62]

This can be seen as a reflection of the hardening of Peto's attitude towards labour over the years, or perhaps as a reversion to the confrontational stance he and Grissell had adopted two decades earlier. In 1859, as a member of the London Master Builders Association, he had become embroiled in another dispute with the building workers, this time over wages and hours. The support he was alleged to have given the employers during the strike that year, caused considerable resentment in the trade union movement, which did its best to discredit his candidature during his election campaigns at Finsbury in 1859 and Bristol six years later. The significant legislation on industrial injuries was not enacted until after Peto retired.

Peto told the Select Committee that there was 'very severe discipline' on all his works. Any ganger who supplied 'tommy' or withheld wages, was instantly dismissed, as was any labourer who failed to report regularly for work without good cause. The sale of alcohol on site was strictly forbidden because, in his view, drunkenness was 'the parent of all accidents'. But he did not stop any labourer from bringing his own supply of beer, insisting that 'a man has a right to bring a gallon with him if he likes in the morning', which his wife could supplement when she took him his midday meal. This policy appears to have kept drunkenness within

bounds, although even his men were prone to bouts of drinking, usually worst after they had been paid at the end of the week.

Peto's claim that his system produced an orderly workforce is borne out by the comparatively few recorded instances of serious misconduct on his works. He did admit that fracas had broken out on occasions between English and Irish navvies on the Great Western line, whilst a serious riot at Folkestone during the construction of the South Eastern line may have involved some of his men. There was also disorder at Brockenhurst during the final stages of the construction of the Southampton & Dorchester Railway, but that was of course on an entirely subcontracted line. In some cases the railway workers may have been wrongly blamed, as for instance, when a spate of rick burning occurred during the construction of the Norwich–Brandon line. There was also trouble on occasions abroad, Peto's agent in Denmark being obliged to repatriate some of the labourers he had brought over from England, as well as a contingent of Swedes.

The occasion when the conduct of his labourers proved personally most embarrassing for Peto occurred some distance from their place of work. On the eve of the poll at Norwich in 1847, a sizable contingent of navvies from the Ely & Peterborough line arrived in the city, ostensibly to demonstrate support for their employer. Things got out of hand when they met a procession of supporters of one of the rival candidates, the Radical J. H. Parry, who were marching through the city headed by a band. In the ensuing fight some windows were broken and the band had some of their instruments damaged, one of the navvies managing to jump through the big drum! After a number of arrests had been made, the main body of navvies was escorted back to the railway station and put on a train for Ely. Although violence was common enough at election times in those days, questions must have been asked as to how such a large body of men came to be assembled so far from their place of work. If, as seems very likely, the demonstration had been planned, what had been intended as a display of working-class support for the Liberal candidate, ended up as a exhibition of navvy brutality. The incident did not prevent Peto topping the poll the next day, although he was subsequently obliged to pay for the damage to the musical instruments. The offending navvies could have included some of those who, according to Peto in his evidence to the Select Committee, had earlier been described, by no less a figure than the Dean of Ely, as an example to the whole district.

His examination by the Select Committee had been in every way a triumph for Peto. He had provided all the evidence needed to convince it, and the world at large, that he was not only a good employer but a paragon to be emulated by his fellow railway contractors. The performance was so masterful and the self-publicity so succinctly expressed, that it attracted the adulation of some of his contemporaries and established a reputation still largely intact today. In 1851 *The Illustrated London News* lauded his 'ever-vigilant supervision of the welfare of the people committed to his care' and more recently J. C. G. Binfield in *The biographical dictionary of British radicals*, after referring to his firm's ruthless attitude towards the building strikers, insists 'in Peto's particular sphere of railway

contracting, there was a genuine concern, both effective and paternalistic, for the navvies' welfare'. This echoes the view expressed earlier by Terry Coleman in *The Railway Navvies.* Whilst there is no evidence to contradict the view, expressed by Binfield that Peto was 'benevolent within contemporary limits', it seems unlikely that he was motivated entirely by altruism. The primary aim of the organisation he set up was efficiency of operation, and a co-operative workforce was essential in achieving that end. Most of the measures he introduced for the benefit of the labourers, such as proper housing, pastoral care and regular pay, contributed to achieving that objective. That they appear in large measure to have succeeded, is certainly very much to his credit and his ability to make substantial profits during the Mania years, justification in itself for his policy; although it is not certain to what extent he persisted with it in the less affluent years that followed. Even in the 'forties the system he employed had been authoritarian as well as efficient and may have been the main reason for him being able to boast to the 1846 Committee that he found his navvies 'easy to control.'

TEN

Achievements and Failings

WHEN he began as a railway contractor, Peto was possibly unique, in bringing to the new business the resources and expertise of a large, well-established building firm. He contrasted markedly with other contractors at the time, who were mostly small-scale operators, many with very humble origins. In contrast to them, he was able from the start to undertake the largest works on offer and as contracts got larger, had no difficulty in expanding his activities accordingly. From 1834 to his final retirement in 1875, Peto undertook the construction of more than 850 miles of railway in Britain and a further 1,500 miles abroad. Although this was considerably less than Brassey – who built more than 2,000 miles at home and nearly 4,500 miles abroad – about a tenth of Brassey's work at home and more than a quarter abroad were joint enterprises with Peto. Besides being vastly important in his own right, Peto was also part of the greatest force in British railway building in the middle of the nineteenth century.

CONTEMPORARY ACCLAIM

PETO's contemporaries were staggered by the physical achievement and the organisational skill that had made this possible. In 1851 John Francis in his *History of the English Railway* thought Peto worthy of a whole chapter, putting him on a par with engineers like Brunel the Stephensons, and the then disgraced Railway King, George Hudson. Of Peto he wrote:

> In forming a tenth part of the the railways of the United Kingdom, he has converted

an ample into a princely fortune ... who much less than a quarter of a century previous, held the trowel or hammered the nail, might have been heard seconding the address on a Queen's speech, listened to by gentlemen and applauded by scholars.[63]

In the same year *The Illustrated London News* extolled Peto's virtues as an employer, saying:

> The system pursued by him may be described ... as combining discipline, personal freedom, moral admonition reduced to practice, and a total avoidance of ostentatious purism ... It is not however to be supposed that his effective control over his dependents is owing to any of the brusqueness of mien and dominating manner, which some masters believe to be essential in their ascendency. On the contrary his demeanour to his inferiors is as unobtrusive and considerate as his bearing to his equals is courteous, polished and self possessed.'[64]

This is to completely overlook his handling of the dispute with his employees ten years earlier.

Equally favourable comment came from the City financial journalist, Edward Pugh Rowsell, who described Peto in an article in *Bentley's Miscellany* in 1855, as 'a highly intelligent, very energetic, and remarkably clear-headed man', admiring the fortune he had made for himself and the way in which he used this to promote worthwhile new enterprises:

> A perfect mountain of capital employed unceasingly with the utmost boldness, at the same time almost unerring sagacity.

We have successful gamblers, and we have lucky speculators on the Stock Exchange, but the mighty contractor who invests tens of thousands in a gigantic operation, only does so upon a basis which can scarcely fail, and calculations which hardly admit of mistake. And when, carefully advancing his operations, so that instead of clashing and interfering with one another, he advances more and more freely, not into the arena of tremendous speculation, but into the legitimate field of sound but bold and vigorous enterprise.

THE CRITICS

THE earliest public criticism of Peto's business methods came from the anonymous author of *Petovia*, who only two years after Rowsell's eulogy, said:

I consider him the railway gambler par excellence – the successor of the notorious Hudson– in the wholesale extent of his undertakings, the ingenuity with which he cooks prospectuses, in the magnitude of his audacity and the success with which he angles for dupes. But the pitcher which often goes to the well is tolerably sure to be broken in the long run and the real friends of the Baronet are they who caution against the madness of playing double or quits.

The events of 1866 were to prove this to have been good advice and give substance to much of the criticism. At that time some of the most outspoken attacks on Peto are to be found in the financial columns of *The Times* and *The Economist*. More recently this view has been restated by P. L. Cottrell in his authoritative account of Peto in the *Dictionary of Business Biography*, whilst Jack Simmons has compared him very unfavourably with Brassey, writing in his preface to the reprint of Helps's *Life*:

There was an absolute integrity in Brassey's character that gave him the trust in an extraordinary degree of all those who employed him: businessmen in his own country, rulers and politicians abroad … Beside him Peto appears to us unctuous, shifty, brash; a more complex man, cleverer, more showy and perhaps more interesting, but in the end entirely dwarfed by Brassey's grandeur.

ACHIEVEMENTS

PETO was in the vanguard of a new breed of large-scale railway contractors, who would come to dominate the industry before the end of the nineteenth century. Like Brassey he was able to execute the major works he undertook quickly, efficiently and with a few exceptions to a high standard. They contrasted markedly with most of the early railway builders, who operated on a much smaller scale and were almost always the total subordinates of the companies that employed them. Peto also had the advantage over many other railway contractors in building most of his English lines in the eastern and southern counties, where there were not many severe physical difficulties. This applied less abroad, however, especially in Canada and Norway.

Most of the complaints about the quality of the construction Peto carried out, came from those critical of his involvement in other aspects of a company's affairs. There are only two instances of a railway being handed over by him in an unsatisfactory state: the Portland extension of the Grand Trunk Railway of Canada, which was a much larger task than he had been led to believe, and the Yarmouth–Norwich line, where he appears to have used defective materials. On the other hand his last two financially disastrous contracts, the London, Chatham & Dover and the Cornwall Minerals, were both built to acceptable standards.

Peto's ability to sustain construction during periods of economic depression and on occasions to facilitate the commencement of works before the onset of boom conditions, resulted in more railways being built and earlier, than would have been the case without his involvement. Some worthwhile projects, notably the City Lines, which he ruined himself building against the odds for the London, Chatham and Dover company, might never have been constructed without his involvement. Other speculations were of less permanent value, however, proving millstones around the necks of the companies saddled with their maintenance, such as the East Suffolk and the Severn Valley. The vision shown in carrying British contracting to the corners of the Earth, although not his alone, and certainly not universally successful, paved the way for large-scale British investment in railways abroad.

Peto not only built railways on a world scale, by the end of his career he had become one of the first truly international entrepreneurs. This was only made possible by the transport revolution he himself had played a part in creating, and Britain's adoption of free-trade economics, which, as a Liberal politician, he had enthusiastically endorsed. Although the commercial world in general benefited from the works he executed, many of the non-railway projects he backed were ill judged: the steamship service to Denmark was not justified by the potential traffic, the hotel at Colchester was too grandiose and the housing which he and Brassey erected for commuters at Southend predated long-distance commuting by several decades. On the other hand London's Victoria Docks and the founding of modern Lowestoft were positive achievements.

As far as Peto's railway promotions in Britain went, like the private speculations, he had a propensity to back losers. This was always on the cards, as these mostly involved weak companies, which had only turned to him for assistance when they found they could not raise the necessary capital on the open market. His primary objective in supporting these companies appears to have been to obtain a new contract on his own terms, although the possibility of disposing of the assets at personal profit after completion, would also have been included in his calculations. This sometimes happened, as with the West End of London & Crystal Palace and East Suffolk schemes, but on other occasions left him with expensive long-term liabilities. When the financial involvement was not excessive, this was probably acceptable but

unfortunately for Peto, the two schemes which most stretched his resources, the Grand Trunk of Canada and the London Chatham & Dover, were also the biggest loss makers.[65]

THE CASE AGAINST PETO

ONE of the major charges made by Peto's critics was that he grossly overcharged for the railways he promoted and built. In *Petovia* it is claimed that leased lines like the London, Tilbury & Southend, were built for about half the price paid by the company and that the money Peto was able to invest as a result more than covered the interest he paid to the company's shareholders. This assumed, however, that all the capital needed could be raised on the open market, which was often not the case, and that the lease did not become a costly liability. In the absence of competing tenders, Peto was almost certainly able to charge more than the going rate for the leased lines, but he would have seen this as an insurance policy against the long-term liabilities. He may indeed have expected that some of these railways would not be profitable in the short to medium term and, in the case of the East Suffolk, admitted this to the shareholders, although only after they had paid him to build it.

Peto appears to have worked on the general assumption that in the comparatively easy terrain in which he was carrying out most of his contracts, he could construct a railway for £10,000 per mile and earn a minimum of 10% profit.[66] With the exception of the West End of London & Crystal Palace, and of course the Metropolitan Extensions he does not appear to have grossly exceeded this figure, at least in the terms of the initial contract, unless he was obliged to provide significant financial assistance to the company. In the case of the Cornwall Minerals, Peto may have well have underpriced the works, although in the absence of any records of his own, this cannot be verified. It is however certain that he could build the Dereham branch for his own Norfolk company for less than £6,000 per mile and the Chipping Norton line, where he committed his own capital, for even less. Both these works were executed at the end of much bigger contracts, however, which would have significantly reduced the costs, as, besides any surplus materials that might have been available, the necessary plant and workers were already in place. In contrast, the high prices Peto charged for the stations on the Eastern Counties and Great Northern lines, should be related to what were probably reasonable prices for the initial works. There seems little doubt however that there and elsewhere, Peto seized any opportunity that came his way to swell his profits from additional works.

Peto's methods of raising capital to meet his commitments to the railway companies that employed him, are much more suspect than his purely contractual activities. The financial manipulations he resorted to in order to finance the London, Chatham and Dover, whilst not proved to have been criminal, were both devious and recklessly unsound. Some of his malpractices, such as charging for works not yet executed and borrowing on the credit of shares he had not paid for, repeated to a much greater degree the doubtful procedures he had followed with a number of his previous contracts. The only possible justification for such practice in the final analysis, was that the end justified the means. The completion of the railway was certainly Peto's overriding consideration and in furthering that objective, he was clearly prepared to be reckless with both his own and other people's money.

THE SECOND RAILWAY KING?

THE biggest slur on Peto's character in *Petovia* was to equate his business methods with those of George Hudson, the disgraced former 'Railway King', whose dishonest handling of company accounts and illicit share dealing had been exposed earlier. One of Hudson's most notorious practices whilst Chairman of the Eastern Counties Railway, had been to inflate the value of shares by paying dividends out of capital rather than revenue,

and manipulating the account books to achieve that end. Peto could be said to have done a similar thing when he paid investors in the London, Tilbury and Southend company interest on their shares before the line was completed. But he did this openly and legally, meeting the cost, in theory at least, from his own pocket; although in reality the money came from the company via advance payments for the works. The receipt of shares directly from the company without payment, or at discounted rates, was a practice that had been much employed by Hudson. Peto certainly did this with the London, Chatham & Dover shares and was obliged in the end to confess he had made substantial gains on some, if, by no means, all of these. The transactions were part of an overall strategy aimed at raising much-needed capital to allow the ailing company to complete its mammoth undertaking; which also necessitated him accepting large quanties of unsaleable shares. Like Hudson, Peto traded upon the confidence in which he was held by gullible and greedy shareholders, whose venom he also had to face when things went wrong. Unlike Hudson, he did not, however, find himself even temporarily in prison as a result of his alleged debts, although he came quite close to this at one stage and, like the Railway King, had to make substantial reparations later.

Despite these and other parallels, including having acquired their initial wealth through inheritances possibly obtained in questionable circumstances, Peto and Hudson were very different personalities. Hudson was essentially the front man, presiding over boards of directors and entering into the rough and tumble of railway politics; Peto, although very capable of presenting his case in person to a company and addressing a public meeting, was essentially a behind-the-scenes manipulator. This did not prevent him from enjoying the limelight on special occasions, such as at the opening of the railway at Norwich. He also appears to have been fully at ease with his social superiors, even royalty, as he showed in Denmark.

Compared with Hudson, Peto was better educated and in every way more polished, preferring negotiation and subtle persuasion to bluster and bullying, but probably the more devious character as a result. Playing a backstage role also made it easier to place the blame on others if things went wrong; as when he held his Financial Manager responsible for the irregularities connected with the issue of the LC&DR Eastern Extension debentures and blamed the manager of his North of Europe Steam Navigation Company for the failure of that company. He certainly had the ability to impress those with whom he came in contact; one such being Benjamin Moran, the Secretary at the United States Legation in London, who, after dining with him at James McHenry's house in April 1863, described Peto as one of the most impressive people he had met whilst in England.[67]

Despite being active on the railway scene for considerably longer than Hudson, Peto's influence, certainly in Britain, was significantly less. Hudson at the zenith of his power controlled 1,500 miles of line and created one great and lasting system, the Midland, and helped forge another, the North Eastern. Peto never exerted influence over more than half a dozen companies at any one time, which were mostly peripheral to the main network, as is shown by the number that have not survived twentieth-century rationalisation schemes.[68] The significant long-term legacies of his speculative promotions are two major commuter lines in the London area, the London, Chatham & Dover Metropolitan Extensions and the London, Tilbury and Southend.

THE VERDICT

THE ultimate verdict on Peto must be that he was a very successful railway builder and a singularly unsuccessful railway financier. The contracting success was based upon his previous experience in conventional building, the ample resources provided by the large firm he inherited and his adaptability to the demands of the new work. As *Petovia* noted he was 'the right man in the right place.' Starting early in the railway contracting business, he was ideally placed to seize the opportunities offered to large-scale contractors like himself, by the extension of the system during the 1840s and 1850s, and in particular the Railway Mania of 1844–47. Indeed it is no exaggeration to call him a Man of the Mania; for, not only was this the time when he probably made most money, his success then governed his thinking for the rest of his business career.

As he acknowledged in a rare moment of introspection in 1846, Peto had been able to transform himself by that time from a contractor into a fully fledged capitalist. His subsequent failures as railway financier were due to the fundamental weaknesses of the schemes he supported, in particular the GTRC and the LC&DR, the extent to which he overcommitted himself there and elsewhere, and his willingness to accept liabilities that were considerably greater than his own resources. Nevertheless he was able to obtain the additional funds to carry through his ambitious undertakings by reason of his own creditworthiness with private investors, as well as banks and discount houses. It was the failure of one of the latter, as has been seen, that effectively destroyed him.

Peto was, as *Petovia* rightly claimed, 'the railway gambler par excellence', as by encouraging overconfidence in others, he compounded the long-term effects of his own speculations. In Britain and in some cases abroad, this resulted too often in overcapitalised companies, whose lines were expensive to operate on account of the high interest charges they incurred. He can also be seen as responsible for some of the unnecessary proliferation that took place in the system, which was detrimental to the long-term prosperity of the railway industry. This, together with his willingness all too often to bend the rules of financial propriety for his own ends, mean that Peto, despite all his acknowledged achievements, must be regarded as a fundamentally flawed character, who exerted a largely deleterious influence upon the companies he controlled.

ELEVEN

Personal Life

To have accomplished what he did, Peto was obviously blessed with superhuman energy, outstanding organisational ability and the unquestioning self-confidence that was characteristic of so many successful Victorians. He also appears to have enjoyed robust health throughout his life, which enabled him to sustain a hectic lifestyle that included what was for the time, an inordinate amount of travelling.

FAMILY

Despite his total commitment to business, Peto appears to have been a dedicated family man, loyal husband and the devoted father of his large family. He once said that being separated from his family was the worst aspect of the extended periods of travel he was quite often obliged to make. Comparatively little is known about his first wife, Mary, beyond that she was his first cousin and the younger sister of his first partner, Thomas Grissell. He married her in 1830, the year that Henry Peto died and the family ties meant that she would have had an inherited interest in his and her brother's joint business activities. She might indeed have played a significant part in persuading their uncle Henry to bequeath his fortune to them. The marriage lasted twelve years, and produced one son and three daughters but was marred towards its end by Mary's deteriorating health. Suffering almost certainly from tuberculosis, she died at the age of thirty-four, shortly after the birth of a fifth child, who did not survive. It is not possible to know how deeply Peto felt her passing but he was certainly sufficiently moved to publish a valedictory volume shortly afterwards, entitled *Divine Support in Death*. How much, if any, of this essentially theological work, he wrote himself, is very questionable. From the few personal details it contains, it can be inferred, however, that during her life with Peto, Mary found his almost total absorption with business a personal strain, besides being in her view, detrimental to the well-being of their children.

Although Peto's second wife, Sarah, had links with the railway world through her father, Henry Kelsall,[69] she does not appear to have concerned herself greatly with her husband's business affairs. She was twelve years younger than him when he married her in her home town of Rochdale in July 1843, barely a year after Mary's death. The surviving photographs of Sarah in later life show a portly lady with a strong suggestion of the matriarch about her, which, as the mother of eleven, she certainly deserved. These, along with the surviving children from his first marriage, brought Peto's total tally of offspring to fifteen, which was large even by mid-Victorian standards. Of his eight sons, the most noteworthy are Henry, the eldest, who became a barrister and was the author of the 1893 biography of his father; Harold, who was a significant architect before embarking upon an even more successful career as a landscape gardener; and Basil, the youngest, who followed his father into politics becoming a Conservative MP.

RELIGION

Peto's second wife appears to have been the main influence behind his conversion from the Anglicanism in which he had been brought up, to the Baptist Church. Like other dissenters, Baptists were on the radical wing of the religious spectrum and its adherents were looked upon in some quarters as a subversive influence challenging the authority of the Church of England. In fact Peto continued to attend services at his local parish church in Lambeth with his mother for some time after he had joined the Baptists. Although his critics would have seen this as another manifestation of a two-faced personality, it probably signified nothing more than the desire of a broad minded Christian to carry on the family tradition and to please his mother at the same time. Peto certainly never accepted the narrower views of some of his fellow nonconformists, being fond of a tankard of ale and partaking of good wine on occasions. He was also strongly opposed to closed communions making it a condition of his support of the Bloomsbury chapel that it opened its doors to all who wished to attend.

Peto was not slow to embrace his new denomination and soon became the most prominent Baptist layman in the country. He was also for many years its most generous benefactor. In 1847 he built Bloomsbury Baptist Chapel in London *(plate 23b)* at a personal cost of £10,000, leaving a mortgage of £4,000 for the congregation to repay over time.[70] Incidentally, he used some of the Caen stone surplus to the requirements of the house at Somerleyton, for this building. He attended services regularly there with all his family and servants whenever he was in London, right up to the time of his bankruptcy; after which, relations with the elders became strained and he no longer lived in central London. During his affluent years Peto financed the building of other chapels, including contributing towards the cost of converting the former Diorama in Regent's Park into what became known as the West End Chapel.

Peto was naturally attracted by the democratic nature of nonconformity, which shared some characteristics with the liberal free trade philosophy he had so enthusiastically embraced – the absence of ecclesiastical authority enabling laymen to control most aspects of chapel affairs. His wealth, combined with his business experience and natural organisational ability, made him an ideal person to take a prominent part in wider chapel affairs and he was for many years Treasurer of the Baptist Missionary Society. When he became a MP he also took upon himself the task of representing the interests of the wider nonconformist community in Parliament. His efforts here were not always successful, however – his Bill to allow dissenters to be buried in parish churchyards being twice rejected by the House of Commons

RESIDENCES

During the early years of his partnership with Grissell, Peto lived at Albany Terrace, York Road, Lambeth, in a large house that adjoined the firm's workshops, which had been moved there from the City some time before Henry Peto's death. Besides being more spacious, the new site adjoining the Thames was more convenient for the import of timber and other materials, as well as being easily accessible to the West End, where most of the firm's early building work was carried, after the opening of the new Waterloo Bridge. Grissell and their foreman George Allen, also lived in the same street. By the time of the 1841 census Peto headed a household of fourteen persons, which, besides his wife and their four children, included several maiden aunts and four servants. It proved, however, a far from ideal location for Mary when her health deteriorated, being situated on damp former marshland. She died there in 1842.

On his second marriage Peto moved north of the river, to more fashionable Bloomsbury, where the family occupied 47 Russell Square, a substantial property on the east side of the Square. The move was possibly made as much to satisfy Sarah's social aspirations as to distance himself from the memories of his former spouse. As the railway business developed, he no longer needed to live so close to the workshops anyway. Indeed, this required him for several years to adopt a peripatetic lifestyle, leasing properties in Hertfordshire and Norwich, to be near the works that were in hand.

The purchase in 1844 of the Somerleyton estate in Suffolk not only provided Peto and his expanding family, with a second, much larger country residence, it also enabled him to enhance his social standing by joining the landed class (*plate 15a*). The house and estate cost him £86,000. But the expenditure did not end there. Over the next few years, with the help of his sculptor/architect protégé, John Thomas, he proceeded to enlarge and convert what had been a fairly modest Queen Anne house, into a much larger and considerably more flamboyant edifice. Built of brick with extensive use of Caen stone, its architectural style was eclectic, including Jacobean and Italianate features (*plates 14a & b*). How much Peto involved himself in its design is not known, although it is tempting to suggest a link between the neo-Jacobean style employed at Somerleyton and the architecture of Whitmoor House, his Surrey birthplace.

Determined to keep up with the latest trends, Peto instructed Thomas to provide him with a Palm House and a Winter Garden, reminders perhaps of his own involvement with the Palm House at Kew and Paxton's Crystal Palace. He also engaged a leading landscape gardener, W. A. Nesfield, to lay out the grounds at Somerleyton, which included ornamental gardens, the latest in horticultural technology and a large maze. The interior of the house was also extensively altered, one of the new features being a large banqueting hall, the ceiling of which, was decorated in a style similar to that to be seen at some important railway stations; this is not entirely surprising, as Thomas had carried out a number of commissions for railway companies.

Somerleyton was the visual expression of Peto's success in the new industrial world, as well as a statement of the power of the new money it was creating. His enemies saw it a manifestation of all that was bogus about the *nouveau riche*. These sentiments were most vehemently expressed by George Borrow, who had a personal grievance against Peto, for making a substantial profit from the gravel he had extracted from his Oulton Hall estate, when he built the railway to Lowestoft. In *The Romany Rye*, Borrow caricatures Peto as Mr Flaming Flamson, with this thinly veiled description of the activities of his new neighbour at Somerleyton:

> Hurrah for the millionaire! The clown who views the pandemonium of red brick which he has purchased in the neighbourhood of the place of his grand début, in which every species of architecture, Greek, Indian and Chinese, is employed in caricature – who hears of the grand entertainment he gives at Christmas in the principal dining room, the hundred wax candles, the wagon load of plate and the ocean of wine, which forms part of it, and above all the two ostrich poults, one at the head and the other at the foot of the table, exclaims, 'Well! if he a'n't bang up, I don't know who be; why he beats my old lord hollow!' The mechanic in the borough town, who sees him dashing through the streets in an open landau, drawn by four milk-white horses, admidst his attendant out-riders; his wife, a monster of a woman, by his side, stout as the wife of Tamerlane, who weighed twenty stone, and bedizened out like her whole person shone with the jewels of plundered Persia, stares with silent wonder, and at last exclaims 'That's the man for my vote' [71]

Despite the showy aspects of Peto's lifestyle at Somerleyton, there was another side to his activities there. He greatly improved the conditions of his tenants by laying out a model village, with twenty-eight cottages and a school, which were grouped around the green (*plate 13c*). He also rebuilt the parish church and erected a wooden nonconformist chapel in the grounds of the house for the use of those of his own denomination.

Peto had displayed artistic leanings from an early age and throughout his life was a keen watercolourist and sketcher. Far from being the complete philistine his enemies depicted, he displayed genuine taste in the collection he assembled at Somerleyton, commissioning pictures from a number of significant contemporary artists, including Daniel Maclise, Edwin Landseer and Clarkson Stanfield; besides purchasing some fine woodcarvings by James Morris Willcox, to complement work probably by Grinling Gibbons, that was already in the house. Much of the art collection remained in the Hall after Peto sold it to Sir Francis Crossley in 1861.

Although the sale was motivated by the losses sustained on the Grand Trunk Railway of Canada contract, by this time Peto probably no longer felt the need to be a member of the landowning class, whose influence had already declined somewhat with the rise of the new mercantile class, of which he was very much part. Also, despite his rural roots, for most of his adult life Peto had been an essentially urban person. It seems unlikely that he spent a great deal of time on his Suffolk estate anyway, his busy business life being based in London. He handed over the administration of Somerleyton at a fairly early stage to his brother James.

The Bloomsbury house remained Peto's London base throughout the time he owned Somerleyton and nearly all the children of his second marriage were born there. He left it in 1860 to take over the lease of his old partner Thomas Grissell's mansion, no.12, Kensington Palace Gardens. Besides being in a more select quarter of town, this was a larger residence and consequently better suited to the needs of his large family. According to the census of 1861, the household at no.12 comprised twenty-eight persons, including no less than sixteen servants. When its lease expired, he moved to 16 Carlton House Terrace but, despite spending considerable sums of money renovating and redecorating this property, only stayed a few months. In deference to his wife's wishes, he moved back to Kensington Palace Gardens in 1863, to an even larger mansion, that was built for him in the grounds of no.12 at a cost of £50,000 by Lucas Brothers, the firm that had carried out a lot of work for him at Lowestoft (plate 23c). Ironically, when Peto was obliged to dispose of no.12a, following the events of 1866, the buyer was Thomas Lucas.

RETIREMENT

IMMEDIATELY after the bankruptcy proceedings Peto moved his family out of London, leasing Chipstead House, a rambling mansion on the outskirts of Sevenoaks. Leaving them there, he took himself off to Budapest, which was sufficiently distant from Britain to avoid the blaze of publicity which had followed his débâcle. He appears to have lived a quiet life in the Hungarian capital, reflecting no doubt upon his reduced circumstances. His main contact with the outside world during the months he was there was the English Club, where he was able to read the newspapers and enjoy the company of other expatriates. Amongst his visitors whilst in Budapest was his old colleague Thomas Brassey, who was still actively involved in the railway contracting business. During their meeting Peto tried unsuccessfully to interest him in a scheme to improve the navigation of the Danube. One possible reason for Peto's move to Hungary might indeed have been the possibility of obtaining fresh work thereabouts. He had already executed contracts in eastern Europe and the area was ripe for further railway development at the time.

Following his return to England, and whilst attempting to relaunch himself with the Cornwall Minerals Railway, Peto lived for three years at Cowley House on the outskirts of Exeter. After the unsuccessful completion of that project, he took leases first of The Hollands at Yeovil, and then Stargrove, a house on the Berkshire Downs near Newbury, before settling down in the late 'seventies at Eastcote House in Pinner, then on the north western fringes of London, where he lived for ten years (plate 22b). During his time there he managed to maintain an upper-middle-class lifestyle, and was recognised as a significant member of the local community by being appointed a Justice of the Peace.

In his late seventies, Peto made his final move, leasing Blackhurst, a moderate sized villa at Pembury (plate 24a), on the outskirts of Tunbridge Wells, a town he knew quite well, having holidayed there immediately after being first elected to Parliament in 1847. Already in failing health, he died there from liver failure, aged eighty, on 13 November 1889. He was buried in the village churchyard *(plate 24b)*, which was an appropriate resting place for the principal advocate of the right of nonconformists to be buried in parish graveyards. Although *The Times* carried a brief mention of his passing and the local newspapers reported the funeral in some detail, by this time he was a largely forgotten figure; whilst most of his famous contemporaries in the formative years of railway building, such as Brunel, Robert Stephenson and Brassey, were long dead.

Peto left no will and there is no evidence as to the final state of his finances, as the taxation documents in connection with the settlement of his estate have not survived. Although his widow Sarah's estate was only valued at just under £5,000 in 1892, it seems probable that the family assets significantly exceeded this. Even so, these could not have been more than a fraction of Peto's original fortune. This is testimony in part to the risky nature of the business in which he had been engaged for more than forty years but even more to the speculative nature of his own approach to railway building.

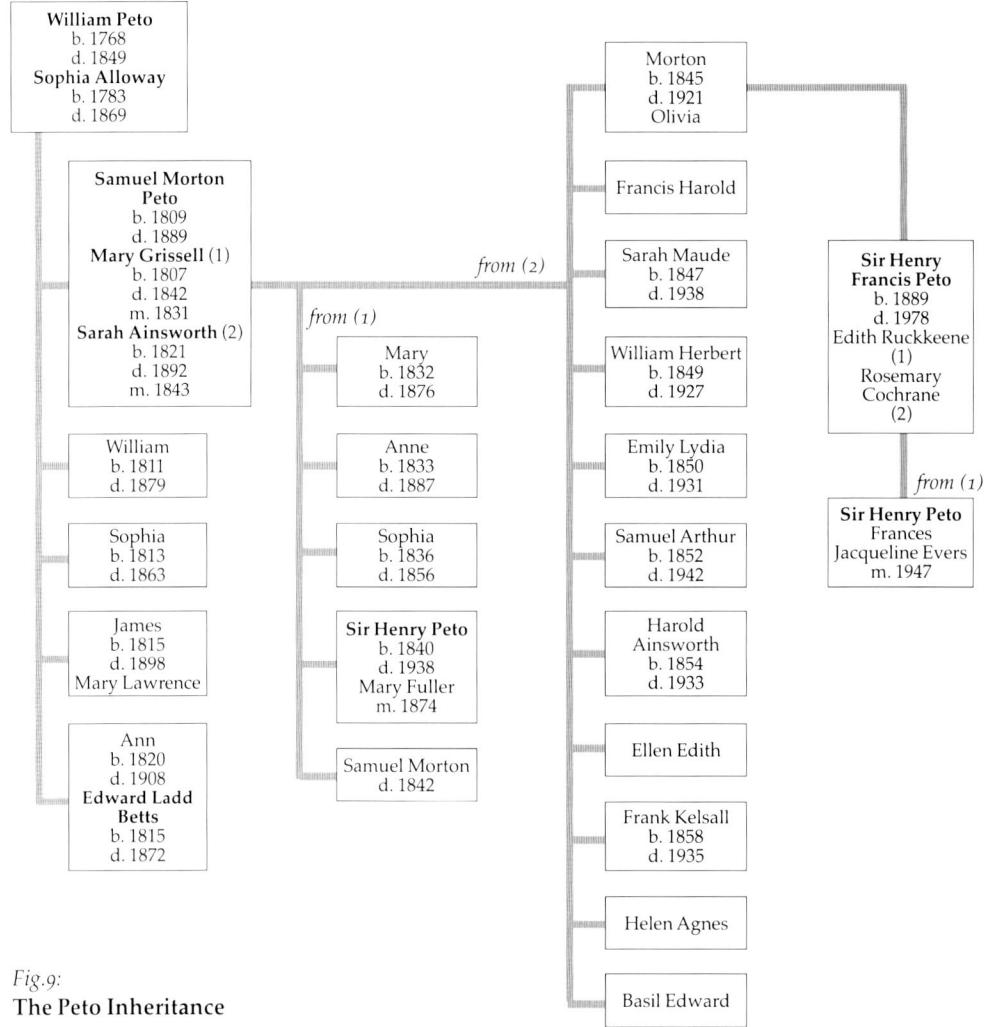

Fig.9:
The Peto Inheritance

Works

Prices quoted are the initial sums agreed with the company for fixed-price contracts – in other cases the final sum received when known

Dates are when the contracts were obtained

WORKS IN BRITAIN

Contracts with Grissell.

1830–31	St John's Church, Paddington and Holy Trinity Church, Chelsea
1832	Hungerford Market, London, £42,000
1833	Birmingham Grammar School, *c.*£30.000
1834	Lyceum Theatre, London
1834	Studley Castle, Warwickshire alterations
1835	St James's Theatre, London, £18,000
1835–39	*Great Western Railway :* Hanwell Viaduct £39,487 Uxbridge Road Bridge £6,570 Iver–Hayes (5 miles) £35,967 Temporary station Paddington Tilehurst–Goring (6 miles) £68,795 Reading station
1836	Reform Club, London
1837	Curzon Street station, Birmingham façade £26, 000 and train shed £10,887
1836–38	Oxford and Cambridge Club
1838	Engine houses and culverts on the Grand Junction Canal
1838–43	*Surrey & Kent Commissioners of Sewers:* Contracts for brick sewers *c.*£3,500
1839	Nelson's Column
1839	Windsor Castle stables £16,900

1839	*London & Birmingham Railway:* Curzon Street Station, Birmingham £26,000
1839–40	*London & Blackwall Railway:* Contracts 3&4 (2 miles) £43,680 Brunswick Wharf buildings £10,482
1840	Pentonville Prison, £84,160
1840	Conservative Club
1840–42	Houses of Parliament
1841	Woolwich dockyard extensions
1842	Thames Tunnel shaft superstructure
1842	Severn Navigation improvements
1842	*South Eastern Railway:* Hythe–Folkestone, Foord Viaduct, Martello Tower Tunnel and adjacent lines (12 miles) *c.*£225,000
1842	*Yarmouth & Norwich Railway:* Yarmouth–Norwich (20 miles) *c.*£200,000
1842–46	*Eastern Counties Railway:* Station alterations on the Colchester line £4,600 Newport (Essex)–Brandon (46miles) Ely–Peterborough (30 miles) Cambridge–St Ives (13 miles) March–St Ives (18 miles) March–Wisbech (8 miles) £1,120,000?
1842	Thames Tunnel engine house
1844	Palm House, Kew
1844	*South Eastern Railway:* Bricklayers Arms branch
1844–6	*Norfolk Railway:* Norwich–Brandon (37 miles) Wymondham–Dereham (12 miles) Dereham–Fakenham (12 miles)

1845	*Lowestoft Railway:* Reedham–Lowestoft (11 miles) £80,000
1845	*Southampton & Dorchester Railway:* Southampton–Dorchester and Poole branch (61 miles) £480,000
1845	*South Eastern Railway:* Widening of Maidstone branch and relaying track from Whitstable to Canterbury
1846?	Olympic Theatre, London

Contracts with Betts

1847	*Birmingham & Oxford Junction Railway:* Fenny Compton–Birmingham (35 miles) *c.*£800,000
1847–48	*Great Northern Railway:* Peterborough–Gainsborough (77 miles) Doncaster–Askern Junction (2 miles) Doncaster–Retford (18 miles)
1848	*East Lincolnshire Railway:* Boston–Louth (33 miles) Contract taken over from Warings
1848	*Mold Railway:* (14 miles) £146,000
1851	*Oxford, Worcester & Wolverhampton Railway:* Completion Wolvercote–Worcester Tipton–Wolverhampton (80 miles) £624,000
1854	*Wimbledon & Croydon Railway:* Wimbledon–Croydon (6 miles) £45,500
1855	*Chipping Norton Railway:* Kingham–Chipping Norton (5 miles) £26,000

Contracts with Betts and Brassey

1851	Victoria Docks, London
1852	*Hereford, Ross & Gloucester Railway:* Grange Court–Hereford (23 miles) £230,000
1852–56	*London, Tilbury & Southend Railway:* Forest Gate Junction–Southend Barking–Gas Factory Junction (50 miles)

Minories Warehouse
Thames Haven branch (4 miles)
c.£700,000

1853	*West End of London & Crystal Palace Railway:* Battersea–Crystal Palace (8 miles) £310,000
1854	*East Suffolk Railway:* Halesworth–Woodbridge and branches (62 miles) £930,000? (excludes Haddiscoe, Beccles & Halesworth Railway section built by Peto earlier)
1858	*Severn Valley Railway:* Hartlebury–Shrewsbury (40 miles) £363,690
1858	*East Kent Railway:* Strood–Bickley (24 miles)

Contract with C.&T. Lucas

| 1853 | *Metropolitan Commisioners of Sewers:* Contract for brick sewers in Southwark *c.*£6,500 |

Contracts with Betts and Crampton

| 1860–64 | *London, Chatham & Dover Railway:* Metropolitan extensions including connecting lines and Eastern Extension to Greenwich (20 miles) *c.*£5¼million |
| 1862 | *Crystal Palace & South London Railway:* Peckham Rye–Crystal Palace (9 miles) £756,000 |

Contract with Betts, Kelk and Warings

| 1865 | *Metropolitan District Railway:* Kensington–Cannon Street (5 miles) £1,700,000 |

Contract on own

| 1872 | *Cornwall Minerals Railway:* Newquay–Fowey with branches (44 miles) £230,000 |

WORKS OVERSEAS

Contracts with Betts

1860   *Algiers & Blida Railway* (32 miles)
       and Algiers harbour
1862   *Great Southern Railway (Argentina):*
       Buenos Aires–Chascomus (72 miles)
       £720,000

Contracts with Betts and Brassey

1851   *Norwegian Grand Trunk Railway:*
       Oslo–Eidsvold (43 miles)
1852   *Paris, Lyons & Mediterranean Railway:*
       Lyon–Avignon (67 miles)
1853   *Royal Danish Railway:*
       Tonning–Flensburg and Rendsburg
       (75 miles) £540,000
1853   *Grand Trunk of Canada Railway:*
       Montreal–Toronto, including Victoria
       Bridge over St Lawrence and lines
       from Quebec to Trois Pistoles
       and Richmond (539 miles)
       c.£6million
1854   *Crimean Railway:*
       Balaclava–near Sebastopol and
       branches (39 miles)
1856   *Kaiserin–Elizabeth Bahn:*
       Linz–Mölk (49 miles)
1859–63 Railways in New South Wales and
       Queensland, including viaduct over
       Nepean river (132 miles)
1861   *Jutland Railway:*
       Randers–Fredericia and branches to
       Hosterbro and Nyborg (270 miles)
1863   *North Schleswig Railway* (70 miles)
1865   *Duneburg & Vitepsk Railway*
       (220 miles)

Contract with Betts, Crampton and Barkley

1863   Ruse–Varna (140 miles)

Sources

BIBLIOGRAPHY

GENERAL

Arnold, A.J. & McCartney, S.M, *George Hudson*, London, 2004

Bailey, B., *George Hudson: the rise and fall of the Railway King*, Stroud, 1995

Bailey, M.R. ed., *Robert Stephenson – the eminent engineer*, Aldershot, 2003

Binfield, J.C.G., 'Peto, Sir Samuel Morton' in *Biographical dictionary of British radicals*, Brighton, 1984

Biographical dictionary of civil engineers in Great Britain and Ireland. Vol.2: 1830–1890. London. 2008

Bowers, F., *A bold experiment: the story of Bloomsbury Central Baptist Church 1848–1999*, London, 1999

Brooke, D, *The railway navvy* 1983

Brooks, E.C., *Sir Samuel Morton Peto Bt.*, Bury St Edmunds, 1996

Chown, J.L., *Sir Samuel Morton Peto*, London, [1943]

Clark, E.F. & Linfold, J., *George Parker Bidder*, Bedford, 1983

Coleman, T., *The railway navvies*, Harmondsworth, 1968

Cottrell, P.L., 'Peto, Sir Samuel' in *Dictionary of business biography Vol.4,* London, 1985

Cox, J.G: 'Railway contractor becomes railway financier: Peto in the 1850s' in *Journal of the Railway & Canal Historical Society* Vol.35, part 1, March 2005

Francis, J., *History of the English railway*, London, 1851

Helps, A., *Life and labours of Mr Brassey; rev. ed.*

with an introduction by Jack Simmons, London, 1969

Ins. Civil Engineers Proceedings, 1889–90

Joby, R.S., *The railway builders*, Newton Abbot, 1983

Lambert, R.S., *The Railway King*, London, 1934

Lewin, H.G., *The Railway Mania and its aftermath*, new ed, Newton Abbot, 1968

Marshall, J., *A biographical dictionary of railway enginers*, Newton Abbot, 1978

Middlemas, R.K., *The railway builders*, London, 1963

Peto, H., *Sir Morton Peto, a memorial sketch*, London, 1893

Peto, S.M., *Divine support in death*, London, 1842

Peto, S.M. 'Navy expenditure' in *House of Commons. Proceedings: speeches of W.S. Lindsay [and others]*, [1862]

Peto, S.M., *Observations on the report of the Defence Commissioners*, London, [1862]

Peto, S.M., *The resources and prospects of America*, London, 1866

Peto, S.M., *Taxation*, London, 1863

Petovia *see* Tooth of the Dragon [Pseud.]

Port, M.H., 'Peto, Sir Samuel Morton' in *Dictionary of National Biography*, Oxford, 2004

Reed, M.C., *Railways in the Victorian economy*, Newton Abbot, 1969

Tooth of the Dragon, *Petovia*, London, 1857

Vaughan, A., *Railwaymen, politics and money*, London, 1997

SPECIFIC CONTRACTS: BRITAIN

Baughan, P.E., *The Chester & Holyhead Railway*, Vol.1, Newton Abbot, 1972

Builder Vol.26, 1868

Christiansen, R., *A regional history of the railways of Great Britain, Vol.7: The West Midlands*, Newton Abbot, 1973

Christiansen, R., *A regional history of the railways of Great Britain, Vol.13 Thames and Severn*, Newton Abbot, 1981

Clinker, R.M., *The railways of Cornwall, 1809–1963*, Dawlish, [1963]

Cottrell, P.L., 'Railway finance and the crisis of 1866' in *Journal of Transport history*, new series 3

Cox, J.G., *Castleman's Corkscrew: the Southampton and Dorchester Railway 1844–1848*, 2nd ed., Southampton, [1985]

Cox, J.G., 'Samuel Morton Peto and the building of the Eastern Counties Railway' in *Journal of the Railway & Canal Historical Society*, vol.33, part 5, 2000

Cox, J.G., 'A direct bounty on extravagant quantities', Peto's Great Northern contracts, 1847–50 in *Journal of the Railway & Canal Historical Society*, vol.34, part 2, 2002

Cottrell, P., *Investment banking in England*, PhD thesis, University of Hull, 1974

Dent, D., *150 years of the Hertford and Ware Railway*, 1993

Doble, E., *History of the Eastern Counties Railway*, PhD thesis, University of London, 1939

Dow, G., *The first railway in Norfolk*, 2nd ed., 1947

The Economist, 14 March, 30 April, 14 June, 7 & 23 July, 28 & 29 August, 1868

Gordon, D.I., *A regional history of the railways of Great Britain, Vol.5: The Eastern Counties*, Newton Abbot, 1968

Gray, A., *The London, Chatham and Dover Railway*, Rainham, 1984

Gray, A., *South Eastern Railway*, Midhurst, 1990

Hart, B., *Folkstone's Railways*, Didcot, 2002

Jenkins, S.C. & Quale, H.I., *The Oxford, Worcester and Wolverhampton Railway*, Blandford, 1977

Joby, R.S., 'Contractor to East Anglia' in *Railway Magazine*, September 1978

Joy, D., *A regional history of the railways of Great Britain, Vol.8: South and West Yorkshire*, Newton Abbot, 1976

Kidner, R.W., *London, Chatham and Dover Railway*, South Godstone, 1952

Leleux, R., *A regional history of the railways of Great Britain Vol.9: The East Midlands*, Newton Abbot, 1976

Macdermot, E.T., *History of the Great Western Railway*, Vol.1, rev. ed., London, 1964

Macdermot, E.T., *History of the Great Western Railway*, Vol.2; rev. ed., London, 1964

Nebarro, G., *Severn Valley steam*, London, 1971

Peto, S.M., *London, Chatham and Dover Railway: report of the proceedings at a meeting of the constituents of Sir Morton Peto ... held at Bristol, 22nd October, 1866*, [1866]

Thomas, D. St J., *A regional history of the railways of Great Britain, Vol.1: The West Country*, rev. ed, London, 1963

Welch, H.D., *The London, Tilbury and Southend Railway*, South Godstone, 1951

White, H.P., *A regional history of the railways of Great Britain. Vol.2: Southern England*, London, 1961

White, H.P., *A regional history of the railways of Great Britain. Vol.3: Greater London*, London, 1963

Williams, R.A., *The London & South Western Railway. Vol.1*, Newton Abbot, 1968

Wolner, A.H., *Economic geography of the development ... of Lowestoft*, MSc thesis, 1956

OVERSEAS

Atlantic & Great Western Railroad. *Reply to the statements of James McHenry and others in relation to the American management of the A&GWR, together with the fourth annual report ... ten months ending October 31, 1866.* Meadville, Pa., 1866 [copy in Library of Congress, Washington DC]

Calhoun, J.C., *History of the Atlantic & Great Western Railway.*

Cambridge Economic History of Europe, Vol. 6: *The Industrial Revolutions and After. Incomes, Population and Technological Change*, part 1

Currie, A.W,. *The Grand Trunk Railway of Canada.* Toronto, 1957

Day, J.R., *Railways of northern Africa.* 1964

Denmark. Generaldirektorate for Statsbanerne, *De Danske statsbaner, 1847–1947.* Copenhagen, 1947

Harrigan, L.J., *Victorian railways to '62.* 1962

Helps, A., *Life and labours of Mr Brassey*; rev. ed. London, 1969

Herapath's Railway Journal, 1866

Illustrated London News, 1851, 1854

Jenson, J.H. & Rosegger, G., 'British railway builders along the lower Danube, 1856–1869' in *Slavonic and East European Review,* Vol.46, 1968

Kohn, I., *Osterreichisches Eisenbahn Jahrbuch.* 1858

Lowe, J.W,. *British steam locomotive builders,* Cambridge, 1975

O'Dell, A.C., *The Scandinavian world,* London, 1957

Peto, H., *Sir Morton Peto: a memorial sketch,* [1893]

[Peto, S.M.] *Die Verkehrs-Interessen Holsteins, Schleswigs und gegenüber dem Petoichen betrrage.* Kiel, 1863

Poor, H.V,. *Railroads of the United States,* New York, 1876.

Robbins, M., 'The Balaclava Railway' in *Journal of Transport History,* Vol.1, no.1, 1953

Skelton, O.D., 'The railway builders' in Wrong, *op.cit*

Stevens, G.R., *Canadian National Railways.* Vol.1, Toronto, 1960

Thestrup, P., *Danske Kongevogne,* Roskilde, 1992

Thestrup, P., *Dampen binder Danmark sammen 1847–1997,* Vol.1, til 1914.

The Times 1868

Westwood, J.N., *History of Russian railways.* London, 1964

Wright, W.R. *British owned railways in Argentina, their effect on economic nationalism, 1854–1948.* Austin (Texas), 1974

Wrong, G.M. and Leyton, W.H., *Chronicles of Canada,* 1916

ARCHIVES

PUBLIC RECORD OFFICE

Great Western Railway:
   Directors' minutes RAIL 250/2
   Brunel collection RAIL 1149/44
London & Birmingham Railway:
   Engineers' reports RAIL 384/104
   Robert Stephenson award to Peto
   RAIL 410/821
London & Blackwall Railway:
   Directors' minutes RAIL 385/1
South Eastern Railway:
   Directors' minutes RAIL 635/17/18
Eastern Counties Railway:
   United Extension Committee:
   Minutes RAIL 186/69/71
   Peto's accounts RAIL 186/7983
Norfolk Railway:
   Directors minutes RAIL 519/3
Lowestoft Railway & Harbour Company:
   Directors' minutes RAIL 441/1/2
Great Northern Railway:
   Directors' minutes RAIL 236/14/15
   Committee of Works RAIL 236/271
East Lincolnshire Railway:
   Minute book 2 RAIL 177/2
Birmingham & Oxford Junction Railway:
   Directors' minutes RAIL 39/3
   Finance Committee minutes RAIL 39/5
Mold Railway:
   Directors' minutes RAIL 496/1
Oxford, Worcester & Wolverhampton Railway:
   Directors' minutes RAIL 558/1/2
Hereford, Ross & Gloucester Railway:
   Directors' minutes RAIL 302/3
London, Tilbury & Southend Railway:
   Joint Committee minutes RAIL 437/1/2
West End of London & Crystal Palace Railway:
   Directors' minutes RAIL 729/2/3
Wimbledon & Croydon Railway:
   Directors' minutes RAIL 751/2/3
East Suffolk Railway:
   Directors' minutes, and reports to:
   RAIL 182/1/2
Severn Valley Railway:
   Directors' minutes RAIL 606/1/2

East Kent Railway:
  Directors' minutes RAIL 415/6
London, Chatham & Dover Railway:
  Directors' minutes RAIL 415
Crystal Palace & South London Railway:
  Directors' minutes RAIL 146/1/2
Cornwall Minerals Railway:
  Directors' minutes RAIL 133/2

OTHERS

Institution of Civil Engineers:
  Frank Smith collection; files on Grissell, Peto,
  Betts. FSB-004, 023, 062]
Peto family papers
Southampton & Dorchester Railway:
  Minute book Southampton Record Office
  D/Z416
Metropolitan District Railway:
  Directors' minutes London Metropolitan
  Archives

OVERSEAS WORKS

Denmark:
Royal Danish Archives:
  Ministeriet for Offentlige Arbejder Jern Bane
  Sager: Kontrakter med Sir Morton Peto.
  1860–1862
  Letter from S.M. Peto to Minister of the
  Interior 1860
  Contract am Aulog og Draft af Jarnbanen
  Kongn. vigt Denmark, 1861

Canada:
Glyn Mills papers:
  Grand Trunk Railway of Canada contract
  Letter books, 1853

Crimea:
Great Britain War Office:
  Papers PRO/WO28/245
British Library Manuscripts Collection:
  43,255(8)

Personalities

BEATTY, JAMES, 1820–1856
Agent working for Peto 1842–1856.
Contracts: Y&NR, S&DR, Crimean Railway,
also carried out surveys for Peto in Nova Scotia
and New Brunswick.

BETTS, EDWARD LADD, 1815–1872
Brother-in-law and partner of Peto, 1848–1866.
Prior to this had collaborated with him on part
of the SER in 1840s and worked independently
on the Grand Junction, Midland Counties, and
Chester–Holyhead lines.

BIDDER, GEORGE PARKER, 1806–1878
Engineer, railway promoter and
mathematician. Personal friend of Peto from
the early 1840s. Assistant engineer to Robert
Stephenson on L&BR and Y&NR. In charge of
works executed by Peto for ECR and NR, LT&SR,
WEL&CPR and ESR; also Royal Danish and
Norwegian Grand Trunk lines. Promoted
Wimbledon & Croydon Railway. Involved in
some of the improvements to Lowestoft
Harbour and in charge of the construction of
London's Victoria Docks.

BORTHWICK, MICHAEL ANDREWS,
1810–1856
Engineer. Assistant to Robert Stephenson
on ECR.

BRASSEY, THOMAS, 1805–1870
Railway contractor and partner/business
associate of Peto, 1851–1866. Carried out many
works at home and abroad, some with other
partners.

BRUNEL, ISAMBARD KINGDOM, 1806–1859
Engineer and champion of broad gauge
railways. In charge of works executed by Peto
for GWR, B&OJR, OW&WR and HR&GR., also
designed *The Great Eastern* steamship, in which
Peto invested. Many other engineering works.

CRAMPTON, THOMAS RUSSELL, 1816–1888
Locomotive designer and railway contractor.
Partner of Peto for LC&DR Metropolitan
extensions and Varna line. Provided
locomotives for CMR.

EARLE, THOMAS *fl.*1850
Executed part of the Y&NR for Peto; also
subcontractor for Norwegian Grand Trunk
Railway.

CUBITT, JOSEPH, 1811–1872
Engineer. Assisted his father, Sir William
Cubitt, on SER and GNR works executed by Peto.
In charge of LC&DR Metropolitan Extensions.

CUBITT, LEWIS, 1799?–1883
Architect. Designed Royal Victoria Hotel
at Colchester for Peto.

CUBITT, SIR WILLIAM, 1785?–1861
Railway and canal engineer. In charge of works
executed by Peto for SER and GNR.

FOWLER, SIR JOHN, 1817–1898
Engineer. In charge of works executed by Peto
for ELR, OW&WR (succeeded Brunel), SVR and
MDR.

GRISSELL, THOMAS, 1801–1874
Cousin of Peto and partner 1830–1846, took
responsibility for Houses of Parliament contracts
obtained in 1839 and 1850 but after dispute over
remuneration with Barry did not complete all
the work originally contracted for.

HARDWICK, PHILIP, 1792–1870
Architect. Designed Curzon Street station,
Birmingham built by Peto.

HINCKS, SIR FRANCIS, 1807–1885
Prime Minister of Canada 1851–54 involved in
promotion of the GTRC.

HODGES, JAMES, 1814–1879
Agent for Peto on NR, GNR and GTRC.
In charge of construction of Victoria Bridge
over the St Lawrence river near Montreal.

HOWE, JOSEPH, 1804–1873
Politician in Nova Scotia involved in railway
promotion.

HUDSON, GEORGE, 1800–1871
'The Railway King'. Railway promoter and
financier. Chairman of many railway
companies but the ECR was the only one built
by Peto. They met occasionally at Board
meetings during the time Hudson was in
charge (1845–49) Also a director like Peto of the
Clay Cross Company.

JACKSON, SIR WILLIAM, 1805–1876
Director of C&HR and partner of Peto in GTRC
and sometime Chairman of the Clay Cross
Company.

KELK, SIR JOHN, 1816–1886
Contractor. Partner of Peto in MDR.

KENNARD, THOMAS WILLIAM, 1825–1893
English engineer in charge of the construction
of the A&GWR. Peto had known him when he
was a boy.

LLOYD, JOHN HORATIO, 1798–1884
Barrister responsible for Lloyd's bonds used to
raise capital for new railway projects, including
the LC&DR.

LUCAS, CHARLES, 1820–1895
Agent for Peto on the NR. Also carried out much
work for him at Lowestoft and leased the
Somerleyton brickworks. With brother Thomas
(1822–1902) established a large building business
in London in the 1850s which had links with
Peto, and they carried out a drainage scheme
together for the Metropolitan Commissioners
of Sewers. Lucas Brothers completed a short
section of the LC&DR Metropolitan Extensions
after Peto withdrew. Thomas Lucas purchased
Peto's second house in Kensington Palace
Gardens (12a).

McHENRY, JAMES, 1817–1891
[sometimes spelt M'HENRY]
Born in Ireland but grew up in Philadelphia,
Pennsylvania. Established a business in
Liverpool in 1850s importing American
agricultural produce into Britain, which made
him very wealthy. Took over promotion of the
A&GWR in 1858. Lived in London from 1860s
until his death.

MERRETT, GEORGE *fl.*1850
Subcontractor on Y&NR and Norwegian Grand
Trunk Railway.

MOORSOM, WILLIAM SCARTH, 1804?–1863
Engineer. In charge of S&DR.

PAXTON, SIR JOSEPH, 1803–1865
Architect and Garden Designer with railway
interests. Designer of the original Crystal
Palace in Hyde Park and responsible for its
reconstruction at Sydenham. Director of the
WEL&CPR. Peto was acquainted with him before
the Great Exhibition of 1851.

PETO, ANN, 1820–1908
Younger of Morton Peto's two sisters. Married
Edward Ladd Betts in 1843.

PETO, HENRY, 1774–1830
Peto's uncle and proprietor of the building
business Peto inherited with Thomas Grissell.

PETO, SIR HENRY, 1840–1938
Son of Peto by first wife, Mary Grissell.
Author of *Samuel Morton Peto, a memorial sketch*
published in 1893. Accompanied father on his
tour of the United States in 1865. Barrister.

PETO, JAMES, 1815–1898
Younger of Peto's two brothers and Steward of
his Suffolk estates. On the Board of the ESR.

PETO, MARY, 1807–1842
First wife of Peto and sister of Thomas Grissell.

PETO, SARAH AINSWORTH, 1821–1892
Peto's second wife. Daughter of Henry Kelsall,
a successful Lancashire textile manufacturer,
who appears to have had substantial railway
investments.

PETO, WILLIAM, 1768–1849
Peto's father. Tenant farmer in Surrey.

RICARDO, JOHN LEWIS, 1812–1862
Financier and nephew of David Ricardo, the
economist. Chairman of the Norwegian Trunk
Railway, 1851–1862. MP strongly in favour of
free trade.

RONEY, SIR CUSACK PATRICK, 1810–1868
Secretary of the ECR, 1845–51. Manager of GTRC,
1853–60. Director of the LC&DR in 1860s.

ROSS, A. M., 1805–1862
Engineer who assisted Robert Stephenson on
the C&HR and later supervised works on the
GTRC, being personally in charge of the Victoria
Bridge, Montreal.

ROWAN, F. J., 1817–1884
Agent responsible for works on ELR, GTRC and
in Denmark. Qualified civil engineer.

STEPHENSON, GEORGE, 1781–1848
Engineer and 'the father of railways'. Promoter
of the Y&NR with son Robert. Involved in the
establishment of the Clay Cross Company.

STEPHENSON, ROBERT, 1803–1859
Engineer. Son of George Stephenson.
In charge of L&BR (until replaced by Bidder),
Y&NR and ECR works carried out by Peto.
Also Consulting Engineer for Norwegian
Grand Trunk, RDR and GTRC lines.
Co-director with Peto of Clay Cross Company.

THOMAS, JOHN, 1813–1862
Stone mason, sculptor and architect.
Designed Somerleyton Hall for Peto and
Preston Hall near Maidstone for Betts.
Also responsible for station at Somerleyton
and ornamental stonework for the C&HR.

THOMPSON, FRANCIS, 1808–1895
Architect employed by ECR, C&HR and GTRC.
Involved with Sancton Wood in the design of
the stations on the Newport–Brandon extension
of the ECR, including Cambridge; and mainly
responsible for Ely and Peterborough.

TURNER, FREDERICK THOMAS, 1812–1877
Engineer responsible with Joseph Cubitt for
LC&DR Metropolitan Extensions and in charge
of works on CP&SLR.

WILLCOX, WILLIAM, 1830–1889
Agent in Australia for Peto and Brassey.

Chronology

1809   Samuel Morton Peto is born on 4 August at Whitmoor House, Sutton, Surrey, the eldest child of William and Sophia Peto.

1821   Peto becomes a boarder at the Brixton Hill Academy, London.

1824   Peto starts an apprenticeship with his uncle Henry Peto, a leading London builder.

1830   With his cousin, Thomas Grissell, Peto inherits Henry Peto's business. The Grissell & Peto partnership is formed.

1831   Partners build St John's Church, Paddington and Holy Trinity, Chelsea. Peto marries Mary Grissell.

1832   Partners win the contract to rebuild Hungerford Market, London.

1833   Partners are engaged to rebuild the King Edward VI Grammar School in Birmingham.

1834   The Lyceum Theatre, London is built. The firm obtains its first railway contract, for Hanwell Viaduct on the Great Western line.

1836   The Reform Club, London, is built.

1839   The firm completes its GWR contracts. Work commences on Nelson's Column. London & Blackwall contract is obtained.

1840   Work begins upon the first of the Houses of Parliament contracts. Strike of the workers on a number of the firm's building sites in London.

1841   Woolwich Dockyard extension is completed.

1842   Peto becomes involved with George and Robert Stephenson in the promotion and construction of the Yarmouth & Norwich Railway. First contract is obtained from the Eastern Counties Railway. Mary Peto dies.

1843   Peto marries Sarah Kelsall.

1844   Peto purchases Somerleyton Hall in Suffolk. Palm House at Kew is begun

1845   The railway to Norwich is completed for the Eastern Counties and Norfolk companies.

1845   The Southampton & Dorchester line is subcontracted.

1846   The partnership with Grissell is dissolved. Peto continues with most of their outstanding railway contracts. Peto gives evidence to the Parliamentary Select Committee on Railway Labourers.

1847   Peto is elected Liberal MP for Norwich. Builds the railway to Lowestoft and begins improvements to the harbour and town. Major contracts are obtained from the Birmingham & Oxford Junction and Great Northern companies. Peto proceeds to carry these out with Edward Ladd Betts, with whom he carries out all his subsequent works except the Cornwall Minerals line.

1848   Partnership with Betts is formalised and Peto provides finance to enable Betts to complete his contract with the Mold Railway.

1849   Peto gives evidence before Committee of Investigation set up by Eastern

1850   Counties Railway after the resignation of George Hudson.

1850   Peto offers a guarantee of £50,000 for the Great Exhibition and is made a Financial Commisssioner.

1851   Peto & Betts begin work with Thomas Brassey on Victoria Docks, London.
Contract is obtained for the Royal Norwegian Railway, Peto's first overseas undertaking and his first collaboration with Thomas Brassey.
Peto becomes Chairman of Norfolk Railway.
Becomes Chairman of Chester & Holyhead Railway.
Seconds the Address to the Throne, with speech in Parliament advocating free trade.

1852   With Brassey, Peto provides half the capital needed to build the Hereford, Ross & Gloucester Railway.
Peto undertakes to build and lease the London, Tilbury & Southend Railway, jointly with Brassey.
Peto is re-elected MP for Norwich

1853   Work starts on the Royal Danish Railway, a joint undertaking with Brassey.
Peto undertakes to construct the Grand Trunk Railway of Canada, jointly with Brassey and Thomas Jackson.

1854   Peto offers to construct a railway to assist the British Army in the Crimea.
Project carried out with Brassey.

1855   Peto is obliged to resign his seat in House of Commons to undertake the Crimean contract.
Peto is awarded a baronetcy for his services in the Crimea.
Becomes Chairman of the Clay Cross Company.

1859   Peto is elected MP for Finsbury.

1860   Peto moves to No.12 Kensington Palace Gardens.
Supports the London, Chatham & Dover Metropolitan Extensions scheme and obtains contract, with Thomas Crampton.
Further contracts obtained in Denmark

1862   Peto joins James McHenry in promoting the Atlantic & Great Western Railroad and becomes Chairman of London Board.

1863   Peto publishes *Taxation*, in favour of free trade and iron warships.
Attacks Admiralty for unnecessary expenditure.
Moves into 12a Kensington Palace Gardens.

1865   Peto undertakes the first portion of the Metropolitan District Railway, with Kelk and Warings.
Peto is elected MP for Bristol.

1865   Peto visits the United States.

1866   Peto publishes *The Resources and prospects of America*.
Becomes insolvent along with his LC&DR contract partners following the collapse of the financial markets on Black Friday 12 May.

1868   Bankruptcy Court discharges Peto and the other LC&DR contractors.

1868   Peto does not contest General Election; Gladstone and Disraeli pay tribute to him in House of Commons.
Leaves 12a Kensington Palace Gardens, moving family to Chipstead House, near Sevenoaks.

1869?   Proceeds alone to Budapest.

1872   Peto undertakes the Cornwall Minerals Railway contract without a partner.

1875   Peto retires from business.

1877   Peto settles at Eastcote House, Pinner.

1884   Peto moves to Blackhurst House, Pembury near Tunbridge Wells.

1889   Peto dies at Blackhurst on 13 November, aged 80 and is buried in Pembury churchyard.

1892   Peto's wife Sarah dies (see p.106)

1893   Peto's son Henry publishes 'A Memorial Sketch' (see p.9)

REFERENCES AND FOOTNOTES

PREFACE

1. Simmons, J., *The Victorian Railway*, London, 1991, p.115

2. Cox, J.G., *Castleman's Corkscrew: the Southampton and Dorchester Railway 1844–1848*, 2nd ed., 1985

CHAPTER 1

3. Crook, J.M. & Port, M.H., *History of the King's works*, Vol.6, 1973, pp.422–30, 451

CHAPTER 2

4. *Reading Mercury*, 4/4/1840

5. See Chapter 6

6. Peto family papers

7. In 2007, after a number of years in other uses, the building has become a hotel once again.

8. Roscoe, T., *The London and Birmingham Railway*, [1839] pp.161–63

9. L&BR Minutes of Directors, 10/4/1839

10. Clark, E.F. & Linfold, J., *George Parker Bidder*, pp.161–63

CHAPTER 3

11. Board of Trade: *Reports of Inspecting Officers of Railways* PRO/MT29/6

12. ECR Committee of Investigation: *Report to shareholders*, [1849], p.28

13. See Chapter 6

14. See Chapter 8

15. GNR *Minutes of Directors*, 21/12/1848, PRO/RAIL 236/14

16. ibid

17. ELR *Minutes of Directors*, 22/2/1848

18. Peto, H., *Sir Morton Peto: a memorial sketch*, [1893], pp.16–17

19. Election poster in Norwich Reference Library Local Collection

CHAPTER 4

20. After ceasing to be Chairman of the C&HR Peto did undertake the construction of the short branch to Llandudno.

21. Jenkins, S. & Quale, H., *The Oxford, Worcester and Wolverhampton Railway*, 1977, p.25

22. For fuller account of the so-called Battle of Mickleton Tunnel *see* Brooke, D., *The railway navvy*, 1983, p.98–100

23. See Chapter 6

24. Norwich Central Reference Library Local History Collection

CHAPTER 5

25. LT&SR Joint Committee no.1 Minutes, 4/8/1852

26. Pollins, H., Railway contractors and the finance of railway development, *in* Reed, M. (ed.), *Railways in the Victorian economy*, 1969, p.219

27. WEL&CPR. Minutes of Directors 3/3/1854

28. *Record of the proceedings connected with the presentation of a service of plate to Sir S. Morton Peto*, 1860, Copy in Public Record Office LIB(y)1/14x/K1789

29. ibid

30. SVR Minutes of Directors, 30/11/1853

31. *Petovia* 1857

32. ibid

33. ibid

CHAPTER 6

34. *The Economist*, 30/11/1867

35. *The Times*, 30/4/1868

36. ibid

37. *Western Daily Press*, 13/10/1866

38. ibid

39. Peto had not offered himself for re-election at Finsbury in 1865, one reason for this possibly being the criticism he had faced for supporting railway proposals affecting the constituency.

40. *Report of proceedings at a meeting of the constituents of Sir S Morton Peto held at Bristol 22nd October 1866*. Copy in the London School of Economics Library

41. *The Economist*, 13/10/1866

42. ibid

43. ibid

44. Full reports of the Bankruptcy Court proceedings appeared in the *The Times*

45. See Chapter 2
46. Report by James Benham and George Kinnear in Bloomsbury Central Baptist Church archives. *See also* Bowers, F., *A bold experiment*, 1999, p.175–83

CHAPTER 7
47. MDR Board minutes, 23/7/1864
48. William Pease diaries per C.R. Clinker, 13/2/1875

CHAPTER 8
49. *Illustrated London News*, 13/9/1851
50. ibid
51. *Illustrated London News*, 11/11/1854
52. Original letter is in the Royal Danish Archives in Copenhagen.
53. Peto, H: *Sir Morton Peto: a memorial sketch*, 1893, p.41
54. *Illustrated London News*, 9/12/1854
55. See Chapter 2
56. There was a model of the Kadikoi incline in the former Museum of Army Transport at Beverley in East Yorkshire.
57. Original document in British Library archives.
58. See Chapter 5

CHAPTER 9
59. Easted was based at Malvern. The partnership was dissolved in 1848. Institution of Civil Engineers archives.

60. Select Committee on Railway Labourers. Minutes of evidence, 1846, pp.72–82
61. Peto, S.M., *Records of proceedings connected with the presentation of a service of plate ... at the Royal Hotel, Lowestoft ... July 18th 1860*, pp.32–33
62. House of Commons debates, 19 March 1862

CHAPTER 10
63. Francis, J., *A history of the English railway*, Vol.1, 1851, p.226–73
64. *Illustrated London News*, 3/2/1851, p.106
65. Peto may also have incurred losses on the Royal Danish Railway and Victoria Docks contracts, as his partner Brassey refers in his will to liabilities there.
66. Peto told the LT&SR directors during the negotiations over the contract, that he aimed to make a minimum 10% profit on all his works.
67. *Journal of Benjamin Moran, 1857–1865*, ed. S. Wallace & F.E. Gillespie, 1948
68. The lines promoted by Peto that have closed are: the Hereford, Ross & Gloucester, the South London & Crystal Palace, large portions on the Lincolnshire lines and the Severn Valley, although part of this is now a successful heritage line.

CHAPTER 11
69. See Chapter 2
70. Bowers, F., *A bold experiment*, 1999, p.31
71. Borrow, G., *The Romany Rye*, reprinted 1906, p.331–32

INDEX